Reducing the Time from Basic Research to Innovation in the Chemical Sciences

A Workshop Report
to the
Chemical Sciences Roundtable

Chemical Sciences Roundtable
Board on Chemical Sciences and Technology
Division on Earth and Life Studies

NATIONAL RESEARCH COUNCIL
OF THE NATIONAL ACADEMIES

THE NATIONAL ACADEMIES PRESS
Washington, D.C.
www.nap.edu

THE NATIONAL ACADEMIES PRESS   500 Fifth Street, N.W.   Washington, DC 20001

NOTICE: The project that is the subject of this report was approved by the Governing Board of the National Research Council, whose members are drawn from the councils of the National Academy of Sciences, the National Academy of Engineering, and the Institute of Medicine. The members of the committee responsible for the report were chosen for their special competences and with regard for appropriate balance.

This study was supported by the Research Corporation under Grant No. GG0066, the Camille and Henry Dreyfus Foundation under Grant Nos. SG-00-094 and SG-02-025, the National Institutes of Standards and Technology under Grant No. NA1341-01-W-1098, the U.S. Department of Defense under Grant No. MDA-972-01-M-0038, the U.S. Environmental Protection Agency under Grant No. R-82823201, the National Science Foundation under Grant No. CHE-000778, the National Institutes of Health under Contract No. N01-OD-4-2139, and the U.S. Department of Energy under Grant No. DE-FG02-95ER14556. Any opinions, findings, conclusions, or recommendations expressed in this publication are those of the author(s) and do not necessarily reflect the views of the organizations or agencies that provided support for the project.

International Standard Book Number 0-309-08734-1

Additional copies of this report are available from the National Academies Press, 500 Fifth Street, N.W., Lockbox 285, Washington, DC 20055; (800) 624-6242 or (202) 334-3313 (in the Washington metropolitan area); Internet, http://www.nap.edu

Copyright 2003 by the National Academy of Sciences. All rights reserved.

Printed in the United States of America

THE NATIONAL ACADEMIES
Advisers to the Nation on Science, Engineering, and Medicine

The **National Academy of Sciences** is a private, nonprofit, self-perpetuating society of distinguished scholars engaged in scientific and engineering research, dedicated to the furtherance of science and technology and to their use for the general welfare. Upon the authority of the charter granted to it by the Congress in 1863, the Academy has a mandate that requires it to advise the federal government on scientific and technical matters. Dr. Bruce M. Alberts is president of the National Academy of Sciences.

The **National Academy of Engineering** was established in 1964, under the charter of the National Academy of Sciences, as a parallel organization of outstanding engineers. It is autonomous in its administration and in the selection of its members, sharing with the National Academy of Sciences the responsibility for advising the federal government. The National Academy of Engineering also sponsors engineering programs aimed at meeting national needs, encourages education and research, and recognizes the superior achievements of engineers. Dr. Wm. A. Wulf is president of the National Academy of Engineering.

The **Institute of Medicine** was established in 1970 by the National Academy of Sciences to secure the services of eminent members of appropriate professions in the examination of policy matters pertaining to the health of the public. The Institute acts under the responsibility given to the National Academy of Sciences by its congressional charter to be an adviser to the federal government and, upon its own initiative, to identify issues of medical care, research, and education. Dr. Harvey V. Fineberg is president of the Institute of Medicine.

The **National Research Council** was organized by the National Academy of Sciences in 1916 to associate the broad community of science and technology with the Academy's purposes of furthering knowledge and advising the federal government. Functioning in accordance with general policies determined by the Academy, the Council has become the principal operating agency of both the National Academy of Sciences and the National Academy of Engineering in providing services to the government, the public, and the scientific and engineering communities. The Council is administered jointly by both Academies and the Institute of Medicine. Dr. Bruce M. Alberts and Dr. Wm. A. Wulf are chair and vice chair, respectively, of the National Research Council.

www.national-academies.org

CHEMICAL SCIENCES ROUNDTABLE

ALEXIS T. BELL *(Chair)*, University of California, Berkeley
MARY L. MANDICH *(Vice Chair)*, Bell Laboratories
PAUL ANASTAS, Office of Science and Technology Policy
MICHAEL R. BERMAN, Air Force Office of Science Research
MICHELLE V. BUCHANAN, Oak Ridge National Laboratory
LEONARD J. BUCKLEY, Defense Advanced Research Projects Agency
DONALD M. BURLAND, National Science Foundation
THOMAS W. CHAPMAN, National Science Foundation
F. FLEMING CRIM, University of Wisconsin
MICHAEL P. DOYLE, University of Arizona
ARTHUR B. ELLIS, National Science Foundation
BRUCE A. FINLAYSON, University of Washington
JOSEPH S. FRANCISCO, Purdue University
NED D. HEINDEL, Lehigh University
CAROL J. HENRY, American Chemistry Council
MICHAEL J. HOLLAND, Office of Science and Technology Policy
FLINT LEWIS, American Chemical Society
TOBIN J. MARKS, Northwestern University
PARRY M. NORLING, RAND
NANCY L. PARENTEAU, Amaranth Bio, Inc.
ELI M. PEARCE, Polytechnic University
EDWIN P. PRZYBYLOWICZ, Eastman Kodak (retired)
DAVID R. REA, E. I. du Pont de Nemours & Company (retired)
GERALDINE RICHMOND, University of Oregon
MICHAEL E. ROGERS, National Institutes of Health
PETER J. STANG, University of Utah
ELLEN B. STECHEL, Ford Motor Company
WALTER J. STEVENS, U.S. Department of Energy
JEANETTE M. VAN EMON, U.S. Environmental Protection Agency
FRANKIE WOOD-BLACK, ConocoPhillips

Staff

JENNIFER J. JACKIW, Program Officer
DOUGLAS J. RABER, Senior Scholar
SYBIL A. PAIGE, Administrative Associate
DAVID C. RASMUSSEN, Project Assistant
DOROTHY ZOLANDZ, Director, Board on Chemical Sciences and Technology

BOARD ON CHEMICAL SCIENCES AND TECHNOLOGY

ALICE P. GAST *(Co-Chair)*, Massachusetts Institute of Technology
WILLIAM KLEMPERER *(Co-Chair)*, Harvard University
ARTHUR I. BIENENSTOCK, Stanford University
A. WELFORD CASTLEMAN, JR., The Pennsylvania State University
ANDREA W. CHOW, Caliper Technologies Corporation
THOMAS M. CONNELLY, JR., E. I. du Pont de Nemours and Company
JEAN DE GRAEVE, Institut de Pathologie, Liège, Belgium
JOSEPH M. DESIMONE, University of North Carolina, Chapel Hill, and North Carolina State University
CATHERINE C. FENSELAU, University of Maryland, College Park
MARY L. GOOD, University of Arkansas, Little Rock
RICHARD M. GROSS, The Dow Chemical Company
NANCY B. JACKSON, Sandia National Laboratories
SANGTAE KIM, Eli Lilly and Company
THOMAS J. MEYER, Los Alamos National Laboratory
PAUL J. REIDER, Amgen, Inc.
ARNOLD F. STANCELL, Georgia Institute of Technology
ROBERT M. SUSSMAN, Latham & Watkins
JOHN C. TULLY, Yale University
CHI-HUEY WONG, The Scripps Research Institute

Staff

JENNIFER J. JACKIW, Program Officer
CHRISTOPHER K. MURPHY, Program Officer
SYBIL A. PAIGE, Administrative Associate
DOUGLAS J. RABER, Senior Scholar
DAVID C. RASMUSSEN, Project Assistant
ERIC L. SHIPP, Postdoctoral Associate
DOROTHY ZOLANDZ, Director

Preface

The Chemical Sciences Roundtable (CSR) was established in 1997 by the National Research Council (NRC). It provides a science-oriented apolitical forum for leaders in the chemical sciences to discuss chemically related issues affecting government, industry, and universities. Organized by the NRC's Board on Chemical Sciences and Technology, the CSR aims to strengthen the chemical sciences by fostering communication among the people and organizations—spanning industry, government, universities, and professional associations—involved with the chemical enterprise. The CSR does this primarily by organizing workshops that address issues in chemical science and technology that require national attention.

Innovation, the process by which fundamental research becomes a commercial product, is increasingly important in the chemical sciences and is changing the nature of research and development (R&D) efforts in the United States. The workshop "Reducing the Time from Basic Research to Innovation in the Chemical Sciences" was held in response to pressures to speed the R&D process and to rapidly evolving patterns of interaction among industry, academe, and national laboratories. The aim of the workshop was to identify and discuss approaches that might speed the innovation process by which basic research leads to innovation.

The papers in this volume are the authors' own versions of their presentations. The discussion comments were taken from a transcript of the workshop. In accordance with the policies of the CSR, the workshop did not attempt to establish any conclusions or recommendations about needs and future directions, focusing instead on issues identified by the speakers.

Ned D. Heindel and Andrew Kaldor
Workshop Organizers

Acknowledgment of Reviewers

This report has been reviewed in draft form by individuals chosen for their diverse perspectives and technical expertise, in accordance with procedures approved by the National Research Council's Report Review Committee. The purpose of this independent review is to provide candid and critical comments that will assist the institution in making its published report as sound as possible and to ensure that the report meets institutional standards for objectivity, evidence, and responsiveness to the study charge. The review comments and draft manuscript remain confidential to protect the integrity of the deliberative process. We wish to thank the following individuals for their review of this report:

Christopher T. Hill, George Mason University
George E. Keller II, Union Carbide Corporation (retired)
David E. Nikles, University of Alabama
David J. Soderberg, BP Chemicals
Kimberly W. Thomas, Los Alamos National Laboratory
Francis A. Via, Fairfield Resources, Inc.

Although the reviewers listed above have provided many constructive comments and suggestions, they did not see the final draft of the report before its release. The review of this report was overseen by Louis C. Glasgow, E. I. du Pont de Nemours and Company. Appointed by the National Research Council, he was responsible for making certain that an independent examination of this report was carried out in accordance with institutional procedures and that all review comments were carefully considered. Responsibility for the final content of this report rests entirely with the organizers and the institution.

Contents

Summary 1

1 Overview of Trends in Innovation in the Chemical Industry 7
   Richard M. Gross (The Dow Chemical Company)

2 Techniques for Structured Innovation 18
   Allen Clamen (ExxonMobil, retired)

3 The Chemistry Innovation Process: Breakthroughs for Electronics and Photonics 28
   Elsa Reichmanis (Bell Laboratories, Lucent Technologies)

4 DARPA's Approach to Innovation and Its Reflection in Industry 37
   Lawrence H. Dubois (SRI International)

5 Comments on the Advanced Technology Program 49
   Mary L. Good (University of Arkansas, Little Rock)

6 What Have We Learned from Hot Topics? 56
   James R. Heath (University of California, Los Angeles)

7 Industrial Innovation with External R&D Programs 64
   Francis A. Via (Fairfield Resources International)

8 Some New Ideas for Speeding Up the Development of Products from University Research 74
   Kenneth A. Pickar (California Institute of Technology)

9 From Molecules to Materials to Market: A Rational Framework for Products Formulation and Design   83
   Venkat Venkatasubramanian (Purdue University)

10 The Tacit Economics of Modeling: Indifference Curves that Should Defy Indifference   90
   Michael Schrage (Massachusetts Institute of Technology)

11 Successful Innovation Starting in an Academic Environment   99
   Richard K. Koehn (Salus Therapeutics, Inc.)

12 Panel Discussion   106

Appendixes

A Workshop Participants   119
B Biographical Sketches of Workshop Speakers   121
C Origin of and Information on the Chemical Sciences Roundtable   128

Summary[1]

Innovation is the process by which an invention, idea, or concept is converted into a real process, commercial product, or the like. Considerable pressure exists in the commercial sector to shorten the time frame and increase the yield of innovation from basic research, but there is no obvious pathway. Innovation in the chemical sciences—particularly starting at the level of basic research—is complex, often involving multiple interfaces in which chemistry is likened to some other area of science and technology.

This workshop focused on factors such as work processes, systems, and technologies that could enable and accelerate the pace of innovation and increase the yield of major innovations from work in the basic chemical sciences. More specifically, speakers identified teamwork, commitment, standardized portfolio management, clear goals, well-defined milestones, and effective technology transfer as some of the characteristics of innovative institutions and practices. Successful approaches to innovation have taken place in different environments and between different environments—despite infrastructure and cultural differences, both interdisciplinary collaborations and collaborations between industry and academia have proven beneficial for all parties. Funding must also be available to promote innovation at stages of research often ignored.

Through this workshop the chemical sciences community was given the opportunity not only to hear from colleagues who have lengthy experience with innovation but also to pose questions and discuss their own pressing, innovation-related concerns. Short summaries of the workshop presentations are found below; the presentations in their entirety are in the following chapters.

The first speaker at the June 4, 2002, morning session, **Richard M. Gross**, of Dow Chemical

---

[1]David R. Rea of E. I. du Pont de Nemours and Company prepared the presentation summaries of Richard M. Gross, Allen Clamen, Elsa Reichmanis, and Lawrence H. Dubois. The summaries for the presentation of Mary L. Good, James R. Heath, Francis A. Via, and Kenneth A. Pickar were prepared by Ned D. Heindel. Andrew Kaldor wrote the summaries of the presentations by Venkat Venkatasubramanian, Michael Schrage, and Richard K. Koehn.

Company, emphasized the importance of innovation to industrial success, quoting Peter Drucker by stating: "Innovation is the fuel of corporate longevity." Gross pointed out the strong link that exists between science and innovation in the chemical industry. He identified three key macrotrends in innovation—high-throughput research, global teams, and market-driven research. High-throughput research is made possible through the use of computational chemistry and is critically important to accelerating innovation. Global teams reflect how a company does business rather than where the company is located, and their success is dependent on the willingness to share data at all levels—that of individual employees, departments, and the company. Market-driven research is neither applications research nor product tailoring but is a collaborative effort between customer and supplier to target a next-generation need.

Gross believes that people, work processes, and partnerships are critical success factors in innovation. Employees are most productive and happiest when their strengths are matched with the organization's needs. Work processes can be continuously improved by standardizing best practices through the stage-gate process and Six Sigma quality control program. Partnerships have continued to increase at Dow (they have tripled in the past 4 years) and aid speed to market. Gross stressed that speed is essential in the 21st century.

Allen Clamen, retired from ExxonMobil, discussed ExxonMobil's business practices stemming from the belief that a structured innovation program, including a well-organized business portfolio management process, increases the effectiveness of innovation. Rapid innovation requires a corporate commitment to innovation, a culture that encourages risk taking, trained program managers, and a strong link to the market. He cited an Industrial Research Institute study of eight companies that offers specific suggestions to improve the effectiveness of early-stage innovation. Reduced cycle time is a main factor contributing to innovative improvements, and several common approaches used to reduce cycle time were discussed. To achieve major objectives, Clamen encouraged the pursuit of parallel approaches to counter the uncertainty of success and advocated the prioritization of long-range research according to the degree of fit with business strategy, the strength of the supporting science base relative to the industry, the breadth of impact, the existence of multiple approaches, and the expected business value.

Clamen envisioned an ideal environment for innovation. It should have an open, sharing culture and customers who are active in setting product targets. Innovation should be considered vital to the business strategy, and structured processes should be used for both innovation and portfolio management.

Elsa Reichmanis, of Lucent Technologies, illustrated that, through the new functionalities it provides, chemistry is an enabling science for the electronics and photonics industry and specifically for Lucent. Her examples from photonics and lithographic materials design showed the importance of translating long-range product targets to desired molecular characteristics and described the interactions of materials selection, process design, and hardware design. Reichmanis stated that building on previous work is vitally important to keep the time from concept to commercialization to 10 to 12 years. She placed high value on long-range research and believes that experience shows that it is difficult to innovate more quickly than that 10- to 12-year time frame.

Lawrence H. Dubois, of SRI International, concluded the morning session by describing the Defense Advanced Research Project Agency's (DARPA) mission: innovation in support of national security. DARPA aims to solve national problems and enable operational dominance in the battlefield by supporting high-payoff core technologies. It strives to support radically innovative research, a risky strategy requiring strong leadership. Dubois explained that DARPA uses an array of management practices, including highly autonomous program managers, and a blend of multidisciplinary skills to accelerate innovation. Fuel cell development was discussed as an example of how DARPA focuses on an important problem, keeps the end point clearly in mind, and empowers through funding.

The afternoon session began with a presentation by **Mary L. Good**, of the University of Arkansas, who served as the Department of Commerce under secretary for technology during the Clinton administration. Her office had oversight of the Advanced Technology Program (ATP). ATP was established with the mission to overcome the "investment gap" by funding precompetitive early-stage technology and enabling technology research in private companies and universities. The program's existence—and whether the federal government should be involved in private technology development—has always been debated. Good pointed to historic precedents in which the U.S. government has funded the development of private technologies in the past. These included the telegraph, aviation, the Internet, and agricultural technology.

Good maintained that the ATP grants program has had considerable success in product and process development. ATP funds large companies in high-risk development as well as many joint ventures between small business, large businesses, universities, and national laboratories. She noted: "There are lots of success stories, but the real question is (and should be) 'Is the ATP program needed?' not 'Does the program work?'" Good concluded that ATP or a similar program needs to be a strategic piece of the federal government's research portfolio because it provides opportunities for entrepreneurs in any geographic location. It also provides opportunities in areas traditionally neglected by corporate and governmental sponsors. She believes that the strategic federal R&D portfolio must contain a balanced blend of support for fundamental research that is not targeted to any foreseeable commercial use, for applied research, and for technology research.

James R. Heath, of the University of California, Los Angeles, discussed molecular electronics as a prototypical "hot topic," exemplifying the kind of early-stage technology that has extraordinary commercial potential but lacks sufficient certainty to attract development funding. He presented a vision for developing commercially valuable nano-level computers that was in sharp contrast to the reality of limited venture capital or governmental aid available to develop the field.

Heath started by posing questions: "From first principles, what are the physical constraints that define the ultimate size of a computational or memory element?" In more pragmatic phrasing: "Is it feasible to pack the equivalent of 1,000 Pentiums on a single grain of sand?" Heath answered the second question in the affirmative and maintained that an interactive molecular system would eventually be fabricated into a computing machine with molecules acting as the switching components. Despite holding six patents relevant to nanocomputing, one of which was cited by *MIT Technology Review* magazine as being among the new patents most likely to change the world, Heath has problems obtaining financing. He concluded by stating that nanotechnology needs long-term funding toward product development that does not currently exist in the private sector.

Francis A. Via, of Fairfield Resources International, documented his experiences developing industrial collaborations with national laboratories and universities as director of external research at Akzo Nobel. Such collaborations are useful to corporate objectives because they evolve new knowledge and concepts, provide different perspectives on current research, and help accelerate corporate R&D. Via noted that most of industry's external collaborations focus on the upstream discovery stage, although many examples exist in which universities and national laboratories take part in later-stage development. Via provided specific examples of industry-initiated collaborations, which greatly accelerate the development of a concept to a product. Using examples including zeolite catalysts for specialty chemicals, novel chlorination catalysts, and ozone-friendly substitutes for chlorofluorocarbons, Via showed how collaborations not only accelerate discovery but frequently facilitate it. A key organizational principle to foster and accelerate development is to maintain technical capability teams in support of core competencies. These team members can then be assimilated into a focused response team to accelerate new product or process development. Maintaining these capability teams is a growing challenge for R&D management.

Via noted that no matter how useful and productive external liaisons are, they collapse if the company's business commitment to that area ceases. He commented: "Sustaining external collaborations over a long haul is possible only if there is real success early on. It is especially challenging to achieve management support to sustain collaborations in the areas of potential new products for new unproven markets."

Kenneth A. Pickar, from the California Institute of Technology, discussed the growing importance of university-industry collaboration. Corporate central R&D laboratories have been in decline, which has driven industry's need for academic research partners. Such interactions present equally compelling benefits to the university; these include the capture of economic value, an environment attractive to young faculty, and contributions to the economic viability of the local community.

Despite mutual benefits, Pickar noted, "the process of technology commercialization from university research is still characterized—in most universities—by misunderstanding, dysfunction, and lost opportunities. There is a serious cultural impedance mismatch, a lack of trust between the parties." He believes that, for the time for translation of basic research to innovation to be reduced, mechanisms must be found to improve the understanding between university and industry. Pickar described Caltech's unique National Science Foundation (NSF)-funded Entrepreneur Fellows Program, as well as the aggressive pro-patent approach of CalTech's Technology Transfer Office and its "grubstake" program, which uses alumni funding to finance student research with a commercial objective. He concluded with thoughts on specific ways to improve relations between universities and large companies.

The morning session of June 5, 2002, began with a presentation by **Venkat Venkatasubramanian** of Purdue University. His discussion focused on the early innovation process—discovery in the early stages of a project in which the design space is explored for improvements on product formulation based on the original idea. He described product formulation and design as "the systematic identification of the molecular structure or material formulation that would meet a specifically defined need" and explained how that research base is managed using modeling- and knowledge-based techniques.

Venkatasubramanian spoke of three modeling options. The first was a fundamental model depicting the physics and chemistry of the problem, which can be used to predict material properties. The second utilized the experience of formulation scientists, using a rule-based model. The last option was a data-driven approach, in which data are used to make correlations, largely ignoring the physics and chemistry. In his experience, a hybrid framework that mixes all three approaches is the best method of modeling. His lab has developed a single computer program that utilizes all three modeling approaches.

Venkatasubramanian used examples from industry to illustrate his computer program's effectiveness in molecular design. In all three examples the computer program was able to save formulation time, improve the design of models constructed by traditional approaches, or both. He closed his discussion by mentioning that his program has "led to better formulation, new chemistry, and the understanding of the driving forces for all of these problems."

Michael Schrage, of the Massachusetts Institute of Technology, emphasized the importance of human behavior to the economics of and the tradeoffs associated with modeling. Schrage discussed how innovation is based on how people behave around versions of models, rather than solely on the model itself.

In his lecture Schrage identified the "Big Lie of the Information Age." It is usually assumed that people's behavior is directly correlated to the quantity and quality of information available to them, but this is not true. Instead of managing information, Schrage stated that we need to better manage iteration. He described the new wealth as "our ability to iterate and perform more iterations per unit time."

One area that Schrage identified as needing improvement is the communication between scientists and businesspeople. He stated that poor communication yields better models that are less accessible to nonscientists, due to the scientist's inherent interest in the question and the businessperson's interest in only the answer. Additionally, modeling infrastructures would be improved by increased participation by all relevant parties in the modeling process. This is achieved by increasing the usability, increasing the attraction factor, and targeting the evolution of models.

Richard K. Koehn, of Salus Therapeutics, Inc., has extensive experience in managing innovation of technology in universities and bringing it to commercial development. He focused on the intermediate phase of the innovation process, where innovation is transformed into a product with an economic impact. This is the stage at which action by an institution can enhance the economic impact of the discovery. Three factors that significantly increase rates of innovation during the intermediate phase are (1) financial—how much investment is made in the discovery process; (2) administrative—decreasing administrative policies and practices of the university to provide a smooth transition from university to industry; and (3) cultural—how the faculty and corporate institutions see themselves as a larger community.

Koehn then discussed the idea of intramural funding, which he described as "monies mobilized, identified, and deployed by an institution for a specific purpose." He offered examples of intramural funding projects on the university level and methods to make these projects more lucrative. Koehn believes that closer tracking of technology transfer will increase the commercialization of products. In addition, those who invest funds and seek a return on their funds should be involved in management decisions about projects. Lastly, these programs must be responsive to faculty and encourage partnerships between investors and scientists.

1

Overview of Trends in Innovation in the Chemical Industry

Richard M. Gross[1]
The Dow Chemical Company

I would like to begin my remarks with an industrial perspective on the critical role of innovation. Everyone would agree with Peter Drucker's statement: "Innovation is the fuel of corporate longevity. It endows resources with a new capacity to create wealth."

Although there are many forms of innovation in the business world, ranging from new business models to technological innovations, for technological innovations it is important to use the working definition of innovation developed by Joseph Shumpeter as an integral part of his economic model early in the 20th century: "Innovation is the first commercial use of new technology."

One key industrial perspective is that innovation differs from invention. There are many inventions still sitting on the shelf that are not creating value for shareholders, stakeholders, or society. It is with this perspective that I share my thoughts with you.

When thinking about the workshop organizer's request to give an industrial overview and to cover the trends in innovation—barriers, key success factors, and the like—I came to several conclusions. First, I wanted to help set the stage for the talks and discussions later today and tomorrow, while at the same time I wanted to be complimentary to the following talks without needless duplication.

To do this, I decided to use both industry information and specific information from the Dow Chemical Company to illustrate what I see happening in the broader chemical industry. I'll start with some broad trend data on chemical patents and, thus, on innovation in the chemical sciences. Next, I've selected three macrotrends occurring in the chemical industry to speak about, and I'll finish with three critical innovation success factors that I call the three P's—people, processes, and partnerships.

My first source of information about innovation is the Council for Chemical Research (CCR) study completed in the year 2000. There were two main segments of this CCR study. First, a bibliometric study that looked at the strength of the chemical sciences in the United States was undertaken by Fran Narin and his colleagues at CHI Research. Second, Baruch Lev, of the Stern School of Business at

---

[1] R. M. (Rick) Gross is corporate vice president of research and development for the Dow Chemical Company. In this capacity he serves on Dow's Corporate Operating Board, Human Resources Committee, Retirement Board, and Corporate Contributions Committee. Gross was a 1996 recipient of the Dow Genesis Award for Excellence in People Development.

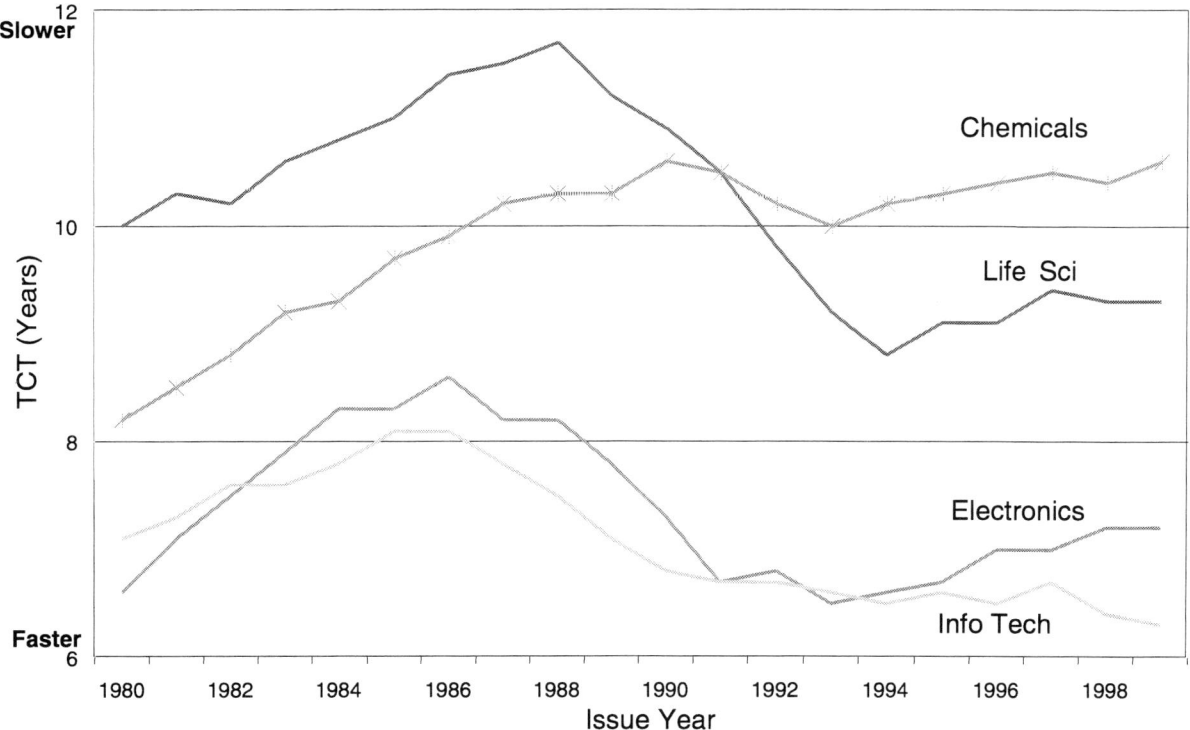

FIGURE 1.1 The technology cycle time (TCT) for U.S. chemical patents is flat, indicating that the industry's rate of innovation has decreased. Courtesy of the Council on Chemical Research, Copyright 2000.

New York University, performed an econometric study on the return on the chemical industry's R&D investment.

Narin's work was based on data from the U.S. patent database, which offers a global view by including the origin of the inventors. It is also dependent on several indices that CHI has developed to look at the impact of patents, the technology cycle time, and the tie to basic science. During the 1980s, the technology cycle time, a measure of the age of the earlier patents cited in a current patent, slowed significantly in the chemical sciences. During the 1990s, it remained flat—a clear indication of the slowdown in the speed at which the industry is innovating (see Figure 1.1).

At the same time, the current impact index of U.S. technology—as measured by citations of U.S. patents by other patents—has strengthened in all areas, including the chemical sciences. A closer look at the data for chemical sciences shows that the U.S. impact is increasing, while Germany's is flat and Japan's is decreasing significantly.

Another index analyzed by CHI was the science linkage, the citation of science publications on patents as opposed to the citation of other patents. The science linkage, the tie to more basic science, is increasing for the chemical sciences over the 15-year study period and is only outpaced by the life sciences. The science linkage data for the chemical sciences again show the United States significantly outpacing Germany and Japan. These trends help frame the importance of the chemical sciences in society and the comparative strength of U.S.-based chemical science innovation, as well as the opportunity in front of us.

The three macrotrends I selected for today's discussion are the impact of high-throughput research, global organizations, and market-driven research on the speed of innovation.

The rapid growth in new high-throughput[2] research tools yields both great benefits and significant pitfalls if not utilized correctly. It is imperative that everyone involved recognizes one important fact. It is simply said: "Since you can go so fast, you better be sure you are going in the right direction." It is also imperative that your objectives and goals are well defined at each level and, most importantly, are understood by everyone. This is where the critical innovation success factors—people and processes—play a key role. People and processes are essential elements of creating alignment throughout an organization and are key to reducing innovation cycle time.

High-throughput research involves how the research is done, not what research is done, and it clearly has the potential to impact the productivity of R&D. Dow became involved in high-throughput research in the late 1990s by partnering with Symyx, a company whose focus is the development of high-speed combinatorial technologies for the discovery of new materials, and bringing Symyx's technology and expertise inside Dow's large corporate R&D organization. One of our goals is to continue to leverage external high-throughput research expertise where appropriate, and we are building a large suite of multidisciplinary tools. The promise of high-throughput research is widely known, and Dow finds it to be a reality. For instance, in the case of a polyolefin catalyst process optimization, over 1,000 high-throughput experiments were run in 6 weeks. There were eight structurally diverse hits, the total time from the first designed experiment to pilot plant runs was less than 5 months, and the cost of the catalyst package was reduced by greater than 75 percent. Dow has additional examples illustrating a 10-fold decrease in cycle time, 3- to 4-fold decrease in personnel costs, and a significant reduction in the scale of reactants used and waste generated. The power is there, but prepared minds need to be thoughtful when setting the research direction before they begin. Without planning, much data can be generated without any knowledge gain.

The second macrotrend of the industry is its move toward global organizations. A global organization is very different from a global company: the term "global company" denotes a location, whereas "global organization" defines a work methodology.

In today's world of specialization, there is a premium paid for being first in the marketplace. To be first in the world with a significant innovation requires global teamwork. Usually the team has a formal structure, but good teamwork among colleagues is just as effective. In fact, good teamwork and collaboration often equate to the ability to communicate effectively. The ability to utilize all of the advanced information technology and communications capability is necessary but not sufficient. Of paramount importance are the ability and willingness to share information freely. Those who do this well will benefit tremendously.

The Wisdom of Teams: Creating the High-Performance Organization by Katenbach and Smith[3] discusses the probability of collaborations as a function of distance. It contains both good and bad news. Unfortunately, once you move past the office or laboratory next door, the rate of collaboration frequency drops off rapidly in the first 90 feet and is only 20 percent of the "next door" collaboration rate. The good news is that past 90 feet the frequency rate changes very little (see Figure 1.2). Although this dataset stops at just over 1 mile, personal experience indicates that with today's information technology and communication tools, collaborations over 1 mile and 1,000 miles are very similar. Low levels of collaboration at a distance are a real barrier to rapid innovation and represent a real opportunity for those who can find avenues for improvement.

---

[2] A high-throughput research tool is any chemical or biological tool set that allows rapid parallel testing of multiple system parameters in a systematic approach. In this instance the author is referring to experimental systems, but these tools can also include computational approaches.

[3] J. R. Katzenbach and D. K. Smith. 1993. *The Wisdom of Teams: Creating the High-Performance Organization.* New York: Harper Business.

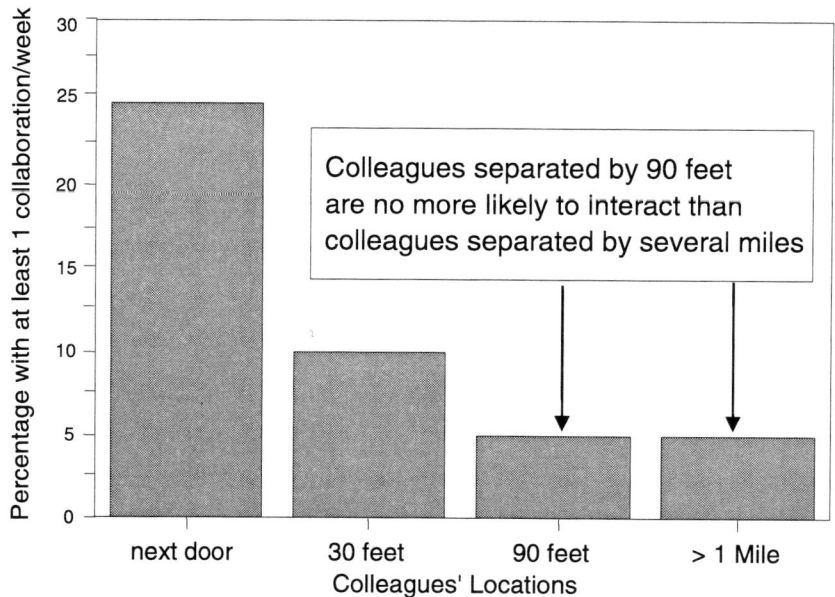

FIGURE 1.2 Although the collaboration rate of employees decreases rapidly over short distances, the rate remains nearly constant for all distances over 90 feet.

At Dow we have standardized our workstations globally. I can go to any of the 50,000 workstations around the world and immediately get to my personalized desktop. The use of NetMeeting, remote network control of experiments, and the sharing of complex spectra and other data globally are standard at Dow, as they are across the chemical industry. All of this has had a large impact on the rate of collaboration.

The last macroindustry trend that will be covered is the increase in market-driven research. Sir Henry Tizzard, a physicist and scientific advisor to Winston Churchill during the war, recognized early on the importance of working on radar technology. This was a remarkable observation that is as important now as it was then. Tizzard said: "The secret of science is to ask the right questions, and it is the choice of the problem more than anything else that makes the man of genius in the scientific world."

To paraphrase Tizzard, the secret of innovation in the chemical industry is to ask the right questions, and it is the choice of the right market opportunity more than anything else that drives the speed of innovation. Identifying the unmet or latent needs in the marketplace and then bringing the full power of basic or fundamental research to bear on the specific opportunity delivers results.

This is not a description of what some people might think of as applications R&D, which is taking an existing product and tailoring it for a specific application. For instance, taking an existing latex formulation and reformulating it for a paper coating opportunity in Europe is an applications R&D activity. Rather, this is a description of the identification of a significant unmet need that requires new materials or a new system to meet the need. Contrary to what some believe, this does not demand a less fundamental approach. In fact, to a large degree, the profitability and sustainability of a company's market position will come from intellectual property and the protection it provides based on fundamentals and on new knowledge derived from basic research.

The market-driven research trend will be illustrated by using some examples from Dow. Although I will move through some detailed information rather quickly, I'll do that in order to paint a larger picture, which is most important.

In the mid-1990s Dow decided to grow its presence in the advanced electronic materials business segment. In an effort to better understand the semiconductor market, one of Dow's top scientists spent 6 months in the marketplace developing the knowledge and understanding required to identify business opportunities where Dow's technical strengths could be leveraged to create a sustainable market position.

There is no doubt that everyone is familiar with Moore's law, the doubling of data density per integrated circuit every 2 years. The performance of integrated circuit devices, historically limited by the characteristics of the transistors, is today limited by the electrical characteristics of the interconnect. The needed improvements in the interconnect performance are achieved with copper and a reduction in the insulator dielectric[4] constant due to the associated reduction in the interconnect capacitance, the cross-talk, and the power consumption.

The scarcity of efficacious insulation candidates prompted the Semiconductor Industry Association to identify the criticality of low-k dielectric material development. Thus, in June 1995, Dow made a business commitment to invent a new material specifically tailored for the interconnect application. Specific performance targets were defined based on interactions with the industry, experience gained through Dow's earlier benzocyclobutene-based systems, finite element analysis of the anticipated interconnect structures, and principles of material sciences.

Molecular modeling was used to predict the dependencies of dielectric constant, mechanical toughness, and thermal stability on the polymer repeating unit structure and cross-link density. The computational chemists worked diligently long before work was done in the laboratory. The computational output was used to focus the targets of the synthesis activities. A synthesis team composed of experts leveraged from throughout the company produced samples from several chemistry families.

SiLK resin is a solution of low molecular weight, aromatic, thermosetting polymer. The polymer's molecular weight and solution concentration were tuned to enable precise and convenient deposition by spin coating, a technique universally used by the industry for the deposition of photoresist materials. After deposition on a wafer, the polymer is thermally cured to an insoluble film that has a high glass transition temperature. The polymer has good mechanical properties at process temperatures, which is required for the application, and it is also resistant to process chemicals.

The most important aspect of this project was the time line. In mid-1996 the specific polymer composition was selected, and in April 1997 Dow publicly announced what became known as SiLK Semiconductor Dielectric. In April 2000, IBM reported the complete integration of the SiLK dielectric and copper wiring and announced its intent to commercially fabricate integrated circuits using SiLK resin.

All of the critical innovation success factors were important in driving this rapid innovation time line. A vast array of external partnerships ranging from universities and institutes around the world to fabrication equipment suppliers and customers were involved. Without the "SiLK network," the project would not have been completed in such a rapid time frame.

The next generation of ultra low dielectric constant material will be a porous SiLK structure. This work is currently being done in partnership with IBM and was started under the National Institute of Standards and Technology Advanced Technology Program. The approach was to template less than 10-nanometer closed pores in the SiLK thermoset matrix. This allows the porous structure to be compatible with the SILK spin-on equipment already owned by integrated circuit fabricators, thus extending the SiLK dielectric through many generations of integrated circuits.

---

[4] The dielectric constant, k, is a measure of the ability of a material to conduct electrons. A low-dielectric constant, or low-k, material is an important part of any electronic circuit because it is used to insulate the copper pathways of the circuit and thereby increases the performance of the device.

The results to date are spectacular. Today we can routinely achieve our goal of closed pores at less than 10 nanometers. To our knowledge, this is a world first in thermoset resins.

Another example that illustrates market-driven research tied to basic research is in the area of polymer light emitting diodes (PLEDs). One of the key attributes of PLEDs is their simple structure compared to that of liquid crystal displays. The chemistry is simple, versatile, and scalable. In short, the chemistry is elegant. Most importantly, the chemistry is tunable across the entire color spectrum. PLEDs were identified as a significant market need by industry leaders. We believed that Dow's expertise could be leveraged to create new knowledge and thus to provide technology options with a proprietary position. Again, this example illustrates the importance of early partnerships. In this case the partnerships were with Richard Friend at Cambridge University and with CD Tech, Inc., also in Cambridge.

This type of innovative partnership occurs across the spectrum of business sectors. Dow's interest in biotechnology began in the Agricultural Sciences business but broadened into nonagricultural applications in the late 1990s. We were interested in using corn plants as production facilities for monoclonal antibodies for use in human therapeutics. The projected growth rate for monoclonal antibody production was high, and a shortfall in production capability was expected—in the year 2005 the shortfall could be as high as 30 percent of demand.

Dow determined that we didn't have the fundamental science base for such research, so we created an alliance with a start-up company, Epicyte Pharmaceuticals, Inc. Our vision is to meet the growing demand for monoclonal antibodies by combining the power of Epicyte's expertise in expressing antibodies in plants with Dow's expertise in the agricultural sciences of corn as well as our overall strength in engineering science and production capability. Combining this new fundamental science with the scientific and engineering strength of Dow Chemical Company clearly was the fastest way to the marketplace.

These three examples illustrate market-driven research that requires substantial fundamental research to generate the new knowledge that can, in turn, generate intellectual property and provide protection in a long-term market position. I would now like to focus on three critical success factors paramount for rapid innovation from basic research to the marketplace—the three P's.

The first critical factor is people. I believe people are the main determinant in successful innovation. The right people are needed at the right time and in the right place. People define the environment, and to have an innovative environment requires the right people. Each person not only possesses a skill set that is typically the focus, but they also have a specific mindset. It is critical to have team members appropriately deployed against the different stages of innovation that match their makeup.

The Myers-Briggs and KAI testing methodologies are tools that describe the personal profile of specific individuals. It is interesting to watch employees relate their test results to what they have felt and experienced on different R&D assignments throughout their careers. In most cases there is an amazingly high degree of correlation. However, sometimes people are attracted to specific types or stages of research based on neither their mindset nor their skill set. They may be attracted by some perception they have about themselves or about that particular part of research.

For instance, there are some people who believe that discovery is more exiting and highly valued than other aspects of research in the development process. In these cases the people are square pegs in round holes. It is therefore very important to move the individual to an area that matches his or her profile, on both skill set and mindset. Proper resource deployment is critical for rapid innovation.

The heart of innovation is ideas. To quote a historic Dow R&D leader, John Grebe: "Ideas are among God's most precious gifts; without them we'd still be living in the Dark Ages. They separate man from all other creatures." He went on to say: "Listen carefully and keep an open mind. Perhaps you can convert a bad idea into a usable one." Grebe clearly understood the power of ideas, and he understood the even greater importance of listening with an open mind.

Fifty years later Arnold Penzias was asked in an interview with *Business Week* what made a top-notch research laboratory. Penzias's answer, concerning the building layout, floored the reporter. Penzias knew the value of ideas and, more importantly, the value of sharing ideas. He knew that no one talks in elevators, so there was no reason to have them. He understood that corridors needed to be long enough to provide opportunities for spontaneous sharing and wide enough that people felt comfortable lingering and building on ideas. Idea sharing is a critical aspect of creating an innovative environment, as is listening to the ideas of others.

Work processes take many forms, from the simplest of structure techniques to the most complex multifunctional processes. They are important, and the more complex the innovation task is, the more important they are. Work processes capture the best practices over time, standardize those practices for everyone on the team, and provide a common language for everyone to use. Additionally, they are key to defining success on multiple levels.

Many people are familiar with work processes, including Bottom Line Innovation, TRIZ, and the various Six Sigma elements. These standardized techniques for work processes have gained a broader use throughout the chemical industry.

At the macro level, much of industry is working with a stage-gate process.[5] Stage-gate processes are business activities, not functional R&D processes. If rigorously used, stage processes are beneficial and can focus employees on the critical scientific technology needs for success. Work processes provide a useful framework to do industry-wide benchmarking to evaluate internal performance versus best-in-class standards. This is useful to identify areas that need improvement and to understand what is both internally and externally possible.

The third critical factor of successful innovation is partnerships. Data show that there are an increasing number of partnerships in the chemical industry and across all industries in the United States.

There are three factors driving partnerships that I would like to mention. The first is industry restructuring. The increased degree of specialization has clearly left many companies without a full hand of cards. Many companies have downsized their R&D organizations or eliminated their central or corporate R&D capabilities entirely. Therefore, necessity has driven a fraction of the increase in partnerships.

Second, understanding the importance of being first in the marketplace causes a company to focus on the speed of getting new products to market. With the rapid rate of change in the marketplace and thus in the industry, it is virtually impossible to have all the right skill sets internally at the right time. Partnerships allow a company to put the required skills together before they are needed, regardless of whether they're internal or external.

Finally, when there is focus on meeting unmet customer needs rather than pushing the technology and science interests of the company, both internal and external people become more willing to utilize all the required skill sets. Dow's partnerships with other laboratories have more than tripled over the past 4 years. I want to close my comments on partnerships by emphasizing the importance of the National Institute of Standards and Technology Advanced Technology Program and others. There are many strong points to these programs, but one that does not get enough recognition is the ability to

---

[5] The stage-gate process is the process by which a new project is evaluated at multiple points in its development. For the project to progress from one stage to the next, it must first pass through a gate—a decision-making point where the choice to continue, kill, hold, or recycle the project must be made. This stage-gate process streamlines the innovation process.

provide a framework for large and small companies to better collaborate and cooperate in the spirit of providing solutions and benefits for society more rapidly.

In closing, I want to end where I started by discussing the three macrotrends in the chemical industry. First, high-throughput technology makes it imperative that your goals and objectives are clearly defined and understood by all. Second, global organizations and global teamwork reflect how we work. Innovation processes must be able to work across distances and do so rapidly. Third, the secret of innovation in the chemical industry is asking the right questions and the choice of the right market opportunity that drives that speed of innovation. In fact, a proprietary and profitable market position will usually only come through new knowledge that comes from basic research.

The critical innovation success factors for the macrotrends above are the three P's: people, processes, and partnerships. People define the innovative environment. Work processes capture best practices and standardize them for all members of the team. Finally, partnerships provide the required knowledge, understanding, and skill sets in real time to really drive rapid innovation—speed counts in the 21st century.

DISCUSSION

Hans Thomann, ExxonMobil: First, do you believe that high-throughput experimentation has had a bigger impact on innovation or invention? In either case, what do you anticipate for the future?

Richard M. Gross: Clearly, what I have tried to capture is that high throughput is unmistakably going to impact invention. We've also been able to take the output of high-throughput research to the marketplace, so the impact on innovation is definitely there.

For the future we must have prepared minds. High throughput is not about what work we do; it is about how we do the work. It needs prepared minds that are skillful and knowledgeable in the area to guide it and to set goals and objectives. Without a thoughtful approach, you can just be very busy generating a lot of data without making any progress on new knowledge. I'm hopeful for the future.

Robert A. Beyerlein, National Institute of Standards and Technology: First, you mentioned that with restructuring activities going on in many companies, they might not have the full resources needed. If the company's science base and resources are not complete for the job at hand, how do you address that? Also, what is the process of implementing innovation that allows the company to focus on unmet market needs? I'm curious about how you identify those unmet market needs.

Richard M. Gross: Let me take these questions in reverse beginning with unmet market needs. The first thing is to work with the marketplace, not specific customers. Customers tend to be focused on their needs, not necessarily the broader market needs. It is beneficial to work with a large array of customers or players in the market, so the entire picture can be seen.

Second, we have found it extremely effective to have one of our most senior scientists involved in that activity. This eliminates the hands-off, nonscientific minds coming back and trying to describe things to the scientific mind. Having a senior scientist involved also puts someone out there who understands the possibilities of science. They then frame and see the problems differently. Having scientific expertise in the marketplace and working with the broader marketplace instead of a specific customer is key to identifying unmet market needs.

The first question was how to handle a lack of scientific resources. We are blessed in my company because we still have a corporate R&D organization that represents 25 percent of our 7,000 employees in R&D. We're very mindful of keeping the basics well tuned.

I think the industry has to pick and choose what expertise it has internally and what expertise it will employ externally. Then it must build those relationships and know the sources it will use for the expertise it lacks.

Dow does that. To anticipate all the needs of the future is impossible. We go around the world to institutes and universities from China to Russia and everywhere in between. We look for the things that add the pieces to the puzzle so we can put the whole picture together.

Michael Schrage, MIT: A few years ago at these kinds of innovation seminars we would have been talking about over-the-wall issues. You take R&D and throw it over the wall to manufacture. You talk about the rise of alliances and the importance of intellectual property.

I'm wondering whether Dow and other companies in the industry have experienced a comparable over-the-wall experience in which you have the scientist working on R&D while the agreements are negotiated over the wall by intellectual property lawyers with boilerplates, who may not necessarily appreciate some of the tradeoffs involved in the collaboration and cooperation issues. How sophisticated have the lawyers needed to become, and is this going to be a topic of greater contention or cooperation with intellectual property (IP)?

Richard M. Gross: Good question. First of all, if there are any IP attorneys in the room, my apologies for my answer. I do not find the IP attorneys to be the most important people. Many companies have created a group to manage their own intellectual assets. At Dow the Intellectual Asset Management Group is part of the R&D function. This group is the interface between the laboratory scientists and the business leaders. The group members understand the business objectives, the business needs, the IP attorneys, and the whole corporation. These folks are well skilled in looking at this strategically.

The way we look at IP is strategically, not as an after-the-fact activity. We spend a lot of time developing and purchasing data-mining technology, so we can look at the topography in order to know where we want to land.

Actually, our intellectual asset management folks use the data-mining technology and other patent-mapping technology to guide our research. This is a strategic thrust for us, and I think it is increasingly becoming a strategic thrust for the industry, not an after-the-fact response type of function.

Henry F. Whalen, Jr., American Chemical Society: You mentioned that you had a partnership with IBM to develop this porous dielectric with less than 10-nanometer closed pores, but you also said that this truly started as an ATP program. I know Mary Good will talk about ATP later, but would you be where you are today if it hadn't been for ATP?

Richard M. Gross: No.

Mary L. Good, University of Arkansas at Little Rock: I wish I had asked that question. Everybody wants to know about processes, but let's go back to your point because it's so very important. What is the industry's thought today on the fact that the enrollment in undergraduate chemistry is going down like a rocket and that graduate enrollment has an increasing number of foreign students? Is that a good thing? What is the industry's thought about that, and what are we going to do about making a difference?

Richard M. Gross: I think industry is unanimously struggling to understand what we can do not just to help at the undergraduate level but to help all kids have a fuller appreciation of science earlier in life. Quite frankly, my biggest concern is not with finding future employees.

My biggest concern is having a science-knowledgeable society to vote. As issues become more technical, I have huge concerns about the people who are voting. I'm concerned about who we're going to hire when I'm retired, but more importantly I'm concerned about the populace. The key from our vantage point is the uplifting of science teachers. It is clear to me that if a teacher has not had an impact by approximately the fifth grade, the student is lost. When you have such a large population of elementary science teachers who are not trained in science and haven't the foggiest idea of what they're teaching, it is little wonder that the kids are not turned on by science.

We're involved with the National Science Resource Center, the partnership between the National Academies and the Smithsonian Institution. Hands-on, inquiry-based science has to be the answer, and it has to be taught by well-informed and well-trained teachers.

Joseph S. Francisco, Purdue University: I was struck by your SiLK example. What I thought was very interesting was the role of computational chemistry in this process. To what extent were the computational chemistry explorations important in terms of reducing the time line, and where do you see it being utilized more in this high-throughput process in the future?

Richard M. Gross: Joe, thanks for the question, because two things come to mind. The answer is that computational chemistry is huge. Whether it's designing a new material for the soles of Nike shoes or whether it's SiLK, the computational chemist could stand here and give you example after example where we've never gone into the laboratory until we've spent hours, weeks, months in front of the workstation.

I have an interesting story about SiLK. When they were doing the computational chemistry, one of the hits they came up with was a material where the published dielectric constant was wrong. The computational chemist brought this forward, and everybody said, "That's not right, because we know what the dielectric constant is. It's published." It turned out the published source was wrong.

Computational chemistry is so important. The three to five families of compounds that were identified by the computational chemists were the only thing that the synthetic chemists worked on. SiLK is indeed out of one of those organizational families. They had material in the marketplace within 6 months of having the original go-ahead for pursuing the thought. This was before the computational work had been done—6 months and they had samples in the marketplace.

There was a second part to this question, Joe?

Joseph S. Francisco: How do you see it going forward in the future?

Richard M. Gross: When we talk about high-throughput research, most people think about combinatorial chemistry in the laboratory. When we think about high-throughput research, we also think about the computational dimension and, most importantly, linking them together.

This is an area where you can share information instantaneously when working in teams globally. The guys in Europe can be working while you are home sleeping and vice versa.

Robert W.R. Humphreys, National Starch and Chemical Company: I applaud you distinguishing between a company that has offices and plants in every part of the world and a company that truly works globally. If the problem were as simple as having computer terminals that speak the same language in every part of the world, we would have solved the problem long ago. What is unique about Dow's training of culturally different people around the world to work together in global teams?

Richard M. Gross: We have a research assignments program in which people hire on to a special assignments program. Most companies have it. We have made that global in recent years. We have new hires that not only do 5- to 6-month assignments around sites in the United States but also go to Europe. This is because we've got 1,400 researchers in Europe. It's our second-largest area in the world.

In their first 2 years, they can have a 6-month assignment in another part of the world. We also have a number of the European analytical chemists come over and do a 6-month assignment in the United States. This program develops networks, which gets back to the people issues that are really at the heart of the matter.

Robert W.R. Humphreys: Is that done through corporate?

Richard M. Gross: Not necessarily. The businesses have a special assignments program as well. In fact, the special assignments program is primarily through the businesses.

Mary L. Mandich, Lucent Technologies: When I saw market-driven research, I immediately thought about financial market drivers. How do you think the forward-looking strategy of corporate-supported research and innovation is affected by market trends and stock prices?

Richard M. Gross: Dow supports our traditional materials that have been around for a very long time. R&D supports them with ever-lower costs and ever-lower resources, because we continually improve our processes and the way we service the industry.

Everyone in the chemical industry is looking for those areas where we can use our capabilities to answer society's problems. Each company has different capabilities and a different focus. Yes, we are a chemicals company, but we're largely a materials company as well. We are trying to identify the growth areas with large enough scope and scale where we can bring our expertise.

I believe that the chemical industry was the first knowledge-based industry. I believe we're still a knowledge-based industry. I believe at the end of the day those in the chemical sciences sell knowledge. I think the challenge is that we all have to look for those marketplaces where we can generate new knowledge to help society as well as be rewarded financially.

2

Techniques for Structured Innovation

Allen Clamen[1]
ExxonMobil (retired)

The highly esteemed management professor Peter Drucker once said, "Success is more likely to result from the systematic pursuit of opportunities than from a flash of genius." I will discuss the systematic pursuit of innovation used at ExxonMobil Chemical Company to increase the yield from basic science to commercialization. Although innovation is thought of as an inherently fuzzy process, my role in the last 5 to 10 years of my 35-year career has been to add a fair amount of structure and discipline to the process of innovation. Some of our best practices will be shared.

The innovation process can be placed into a business context by separating its distinct parts, as shown in Figure 2.1, beginning with the continual flow of ideas from a variety of sources such as customers, academia, and partnerships. Once an idea is selected to continue as a project, it enters our Product Innovation Process or Capital Investment Management Process, which is a stage-gate process that begins with data collection. The project then progresses stage-wise through the complete development of the idea into a commercial product or process. The Portfolio Management Process then analyzes these data so that decisions to progress, accelerate, decelerate, suspend, or terminate can ultimately be made. Of course, these decisions must be made with a clear understanding of the business strategy on which all relevant ideas are based.

The New Concept Development Model (see Figure 2.2), described in *The PDMA ToolBook for New Product Development*[2] which was written by me and several others from different companies, provides a common language and terminology necessary to optimize the front end of innovation. This model was developed under the auspices of the Industrial Research Institute with the hope that we can begin to share best practices in a more understandable way and thus begin to improve upon current state-of-the-art

---

[1]Allen Clamen (now retired) was senior advisor for marketing/technology value creation at ExxonMobil Chemical Company in Houston, Texas. He was responsible for developing effective and efficient processes for idea management, portfolio management, and stage gating of new product development projects. For the past 4 years he has led teams to create value via improved marketing and technology processes.

[2]P. Belliveau, A. Griffin, and S. M. Somermeyer, eds., 2002. *The PDMA ToolBook for New Product Development*, Hoboken, NJ: John Wiley & Sons.

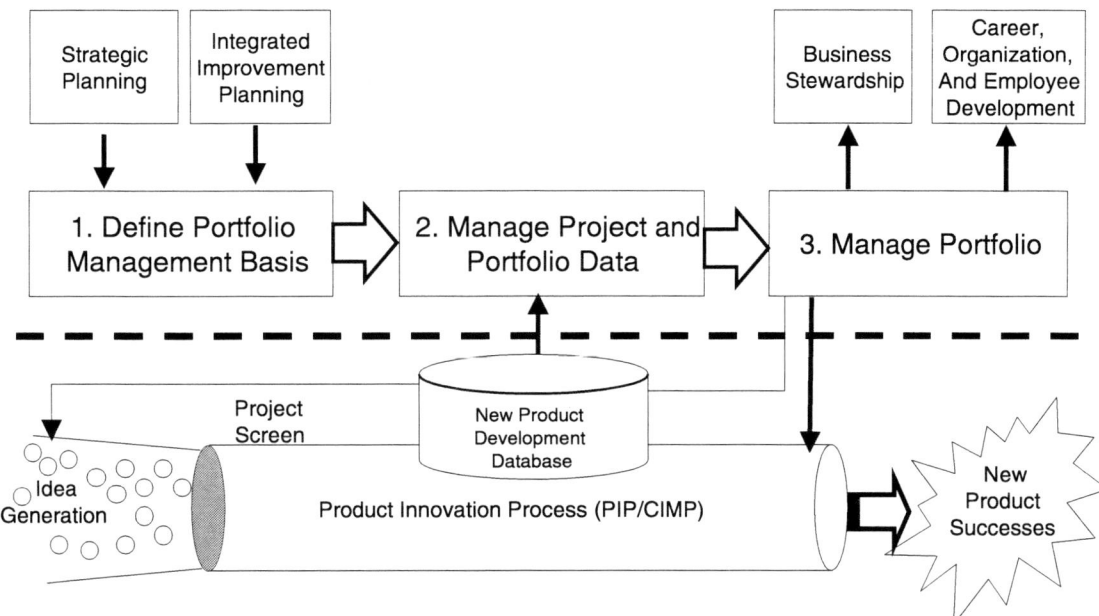

FIGURE 2.1 The innovation process can be encouraged and controlled by breaking it into distinct parts.

innovation practices. The center portion of this model, the engine, represents the leadership, business strategy, and other factors within the company's control. This engine drives the five core front-end elements where activities such as idea generation and opportunity identification take place. The circular shape and curved arrows between these elements indicate that interaction and recycling continuously occur among all five elements. Moreover, these elements are influenced by a number of factors that are largely uncontrollable: the business climate, the economic climate, the customers, and the competitors. From a process standpoint, one can enter the model at any element and ultimately obtain a more detailed understanding of the idea, which is then termed a "concept." At that point, the concept enters the standard stage-gate process for new product or process development.

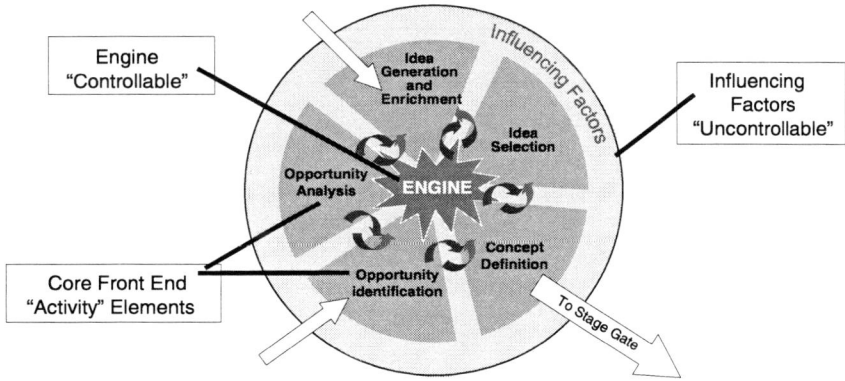

FIGURE 2.2 The New Concept Development Model provides a common language and terminology necessary to optimize the front end of innovation. This figure appears in *The PDMA ToolBook for New Product Development* and is used with the permission of the author.

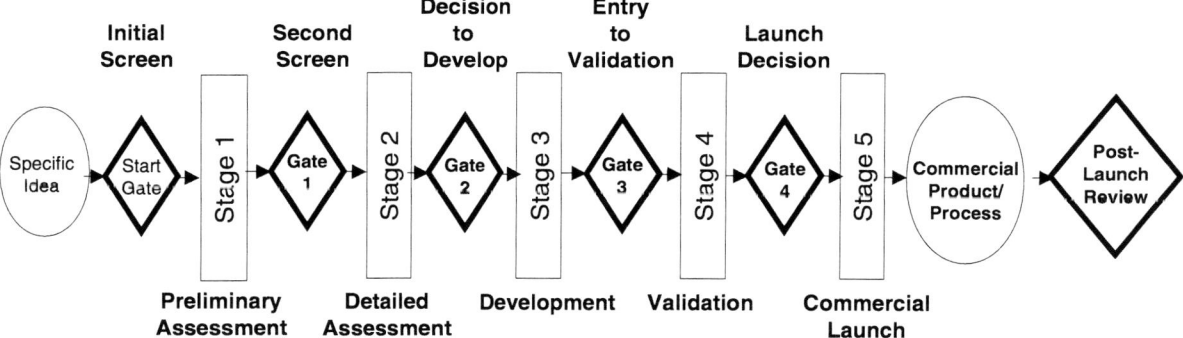

FIGURE 2.3 The stage-gate innovation process.

As mentioned previously, an idea is analyzed, screened, and deemed worthy of further development before it enters our stage-gate process. Figure 2.3 shows the entire outline of the stage-gate process starting from the specific idea (or concept). The first screening is mild, so it does not stifle further idea generation. The preliminary assessment, or Stage 1, of these ideas consists of activities designed to answer a number of basic questions. These include whether the project idea fits with the company's business strategy, whether it can make a profit, whether the market is large enough to justify developing the idea, and whether the idea is technically feasible.

After completion of the activities in Stage 1, the second screen, or Gate 1, determines whether a more detailed assessment involving the customer should be undertaken in Stage 2 to gain a better understanding of the idea's value in the marketplace. The decision to develop or drop the idea is made at Gate 2. The bulk of the project costs are in Stage 3, which is the actual development process. Therefore, the Gate 2 decision involves a major commitment by the company. The decision to commercially launch a product is not made until the customer validates the need for the product and its uniqueness in the market. Once the launch decision is made at Gate 4, the production and marketing plans are finalized, and the product or process is transferred from the project team to the ongoing business. The innovation process is not complete, however, without a postlaunch review. This review takes place in two parts: one soon after the launch and the other about a year or two later. These reviews provide an important assessment of the process and can result in improvements to the process.

Our stage-gate and portfolio management processes are integrated in the stage-gate decision-making process. The first decision is made at each gate meeting by the team of gatekeepers, who are the managers responsible for the resources required to develop the idea. This first decision is based solely on the project's merit, whether or not it has met all the objectives set out at the beginning of that stage. If so, it is deemed ready for resourcing. If not, it is put on hold for further data, killed, or returned to our idea management process for further enrichment or to be combined with other similar ideas. Once an idea is ready for resourcing, the portfolio management team decides on its relative merit, or how it compares with all the other projects in review. In addition, the team considers the balance of projects being resourced (short- versus long-range, how many market segments are represented). This decision-making process must also be integrated with all related business processes such as stewardship and staff development.

By benchmarking some of the most successful companies with respect to innovation, we were able to gather a set of best practices. First, the culture and organizational structure can make or break the innovation process in any company. As a result, it is vital to demonstrate a strong corporate commitment to innovation. It is also critical to maintain a business strategy articulated in as specific terms as possible since not all ideas will survive the entire process and reach the marketplace. Along with this focus, long-term funding must be available to allow the researchers to explore all relevant ideas through the early stages of the innovation process.

Rapid innovation works best when dedicated teams are provided adequate time and responsibility not only to explore new ideas but also to pursue them to the next stage. A culture needs to be created that supports and encourages innovation and the entrepreneurial spirit. Risk taking should be encouraged, especially since errors will inevitably occur and represent an important learning opportunity.

In some companies a top executive is made responsible for growth and innovation. This makes the innovation process more binding so that ideas are retained and used. Also, the formal creation of an "idea manager" provides someone in the process to help coordinate and assist an idea through those initial formative stages. Project management should not be taken too lightly. Projects need to be run by actual project managers, not just the scientist who had the idea or the person who is available because his or her last project has just finished. Peer audits can check the quality of execution of projects. However, even with advances in information technology, it is still considered essential to have everyone involved in the innovation process be in the same place together. The innovation process should also be tracked by a limited number of simple measurable goals, such as the percent revenue from new products, number of patents, and innovation climate.

External participation in projects by academia, other companies, and especially the customer offers a method of validating ideas against current market conditions. This is a difficult part of the business. Also, having a business champion, whether he or she is in management or is a respected scientist who advocates the importance of the project to management, generally correlates with success.

There are two types of issues involved in reducing the cycle time of the innovation process: marketing issues and technical issues. These must be addressed differently. Marketing issues involve the identification of unmet needs by the customer, and technical issues are related to the capability to deliver. Techniques that reduce cycle time include the following:

- Practice openness in partnering with the customer or supplier. The customer ultimately determines the speed of technology acceptance. Success is not the ability to deliver but the customer's ability to use a product in a way that fits the customer's needs. To do this, reliable, up-to-date customer, market, and competitive data must be obtained before the final design along with feedback from the customer. One technique for obtaining valuable feedback from the customer is rapid physical prototyping, which is the process of showing the prototype to the customer as early as possible.

- Measure and communicate cycle time and factors contributing to cycle time reduction. The process of simply measuring cycle time will result in the cycle time being reduced. Instantaneously obtaining information such as project data and progress using web-enabled tools will also support and facilitate the process.

- Formalize project management practices. Formalizing project management practices will reduce cycle time because it ensures strategic and operational alignment across the entire organization. Endorsement of the project by senior management puts emphasis on ensuring that resources are available. The use of cross-functional teams throughout the project is critical to guarantee that everyone is "on-board" when activities are being conducted simultaneously in technology, manufacturing, and marketing. In addition, lessons learned from postlaunch reviews must be applied to avoid making the same mistakes repeatedly. An example of this is overestimation of the size of the market or rate of market penetration.

- Use portfolio analysis techniques to select and progress research and development projects. Research cannot be scheduled, but finances can be budgeted and milestones set. Development, on the other hand, should be done well and as rapidly as possible to maximize profit.

- Share credit for research and development success with the business unit. Better participation and new ideas can occur by providing appropriate incentives and rewards. Methods and practices that motivate all members of the team will foster innovation.

A faster response to the customer's needs will directly impact customer satisfaction. Speed is important not only because it lowers cost and allows the product to be first to market but also because speed yields higher success rates due to the increased likelihood of hitting rapidly changing targets. Of course, less time spent on obtaining results from research and development means that less money is spent, a higher margin is gained, and a longer product life cycle ensues. Overall, reduced cycle time translates to better economics.

The following is a set of management practices that will encourage innovation in any organization. First, a set of clear, consistent, and aggressive goals is needed. Aggressive goals grab the organization's attention and challenge it. Using parallel approaches to pursue a primary objective in a certain area of opportunity will decrease the time spent reaching that objective. This allows the team to work around any problems, because one of the other approaches may have bypassed the problems and progressed beyond that stage. Each approach must be confirmed. This enables customers, competitors, and industry in general to view each approach as a major advance. Although the expense is high, the potential benefits make it worthwhile to move along all of these paths. Unexpected findings should be explored to identify new business opportunities. On the other hand, parallel approaches should not be taken with many different projects. A limited number of objectives and technologies should be selected depending on the size of the organization.

Managers should be prepared to accelerate the most promising projects. Once such a project is identified, as many resources as possible should be given to it. It is also important to keep technical people challenged to maintain the leadership position that the company has in the area of interest.

Communication is always critical to any project. Managers should ensure that their employees talk to each other, the scientists, and the customers. Specifically, the project's scientists should be communicating with the customer's scientists, since they speak the same language.

The following are some criteria found in the literature[3] that generally match those used by ExxonMobil to prioritize long-range research. Of course, a long-range research plan must fit the company's strategy and the strength of the supporting science base relative to the industry. Can the company grow, maintain, and tap into the science that is evolving out of the proposed effort on which the company is about to embark? How wide is the impact of new technology? How many products across the company's slate will be impacted by this new science that is being developed? How robust is the product in the event that there are changes in the business environment? Will the project still succeed? The chances of success are greater if multiple approaches to achieve the goal exist.

Long-range planning also requires estimates of the project's business value. At the very early stages it is difficult to determine what the market share and volume of product sold will be. The company must determine probabilities of both technical and commercial success. These are calculated independently and then combined to finally estimate how successful the project will be.

Innovation is now recognized as being essential to business. While scientists and engineers in an open and sharing culture are the champions of innovation, the customers set the product targets and are the judges of success and failure. Understanding where market needs and technology ultimately intersect

---

[3]G. Tritle, E. Scriven, and A. Fusfeld. 2000. Resolving uncertainty in R&D portfolios. *Research Technology Management* 43:47-50.

as well as involving customers in the early stages of the innovation process will significantly increase the odds of commercial success. In addition, the systematic pursuit of innovative opportunities via the structured processes described above will allow an organization to capture, maintain, and employ best practices to ensure long-term success in innovation.

DISCUSSION

Robert W.R. Humphreys, National Starch and Chemical Company: Allen, you talk about your innovation process, which is not too different from the one we use. When you're doing portfolio management and a lot of projects are coming through, you've got to make decisions. The quality of your portfolio management team is obviously a critical thing or there will be many mistakes that won't be discovered for a long time. How do you choose that team?

Allen Clamen: The quality of the data, the input that goes in, is determined by the effort of the portfolio planner before all the portfolio management team's meetings. These can take place once a month, once a quarter, or as often as is necessary to select projects. All of those data go into the portfolio tool through the database and are collected.

Regarding the selection of team members, it is important to remember that they have the authority to allocate the resources. They would be the managers within each of the functions for which resources are required. It would be the manufacturing manager, marketing manager, and a technology manager. These people would bring with them the resources they need to assess the significance of the data presented. The portfolio tool presents different views, but the portfolio itself is presented for the team to make the decision based on the data, which is hopefully of the right quality, insured by the work that was done up front.

Robert W.R. Humphreys: Is every one of those people market savvy and customer savvy?

Allen Clamen: No, not at all. In fact, there has to be a great deal of trust. The manufacturing manager knows only what he knows relative to what he has learned about the marketplace through his own activities. The marketing manager has an entirely different view, that of the industry at large, the marketplace at large, and many different customers. The manufacturing manager relies to a great extent on the savviness of the marketing manager and his knowledge of that process. Similarly with technology and manufacturing, each of the functions has that degree of knowledge that is not complete for each of the others.

Richard C. Alkire, University of Illinois: To make investments today, your industry and many others use sophisticated economic tools based on assessment of risk for leases of contracts, options, and the like where there's uncertainty in the future. Can those tools be used to deal with research?

Allen Clamen: Yes, the expected value of an idea can indeed be adjusted by its probability of success based on both technology and commercial risk. The technology risk associated with a particular idea can be estimated based on a number of variables that have been validated by past experience.

It is important to learn about the variables as quickly as possible in order to adjust your priorities early based on that full knowledge. One of the things we found to be helpful was a technique called "scenario planning," which looks at many different ways that the world is going to change in the next 5 to 10 years or more. That allows you to say, "Well, if it's going to be a green world, how does that change the business climate and the environment for our products? What does that say we should begin to work on? Here's a project that doesn't take into account this green revolution. Why should we work on that if the environment is looking that way?" Scenario planning allows you to look at several different views of the future world and adjust priorities based on which scenario you believe will take place.

J. Stewart Witzeman, Eastman Chemical Company: I'm intrigued with your comments about portfolio management. How in particular does this portfolio management team function in terms of balancing long-term work versus incremental work, and how do you fit exploratory research and development into the portfolio? In other words, are the managers just managing what they have in front of them today or are they looking at what an ideal portfolio of projects should be for the enterprise?

Allen Clamen: Manufacturing and marketing managers tend to have projects that are a little bit shorter term than your technology manager. They report to a business manager, who is responsible for long-term prospects as well as the long-term success of his enterprise. The business manager must be cognizant that a company can't have 90 percent of its activity in the short term or it will, in the long term, be unsuccessful. That is viewed at each portfolio management team meeting.

We look at the percentage of projects in each time frame and require 30 percent of the projects to be long term as part of the business strategy. Thirty percent may not be enough to some companies, but it probably is enough in the case of a chemical company. It is important to recognize where you are relative to your overall portfolio. This allows you to continue to progress those projects that are longer range and have less support from the team's business people who look at the shorter term.

Michael Schrage, Massachusetts Institute of Technology: It has been my experience working for a fairly broad variety of rather large organizations that the drivers for the bulk of these issues tend to be two words that weren't mentioned at all: politics and the allocation of overhead. Could you respond as to the nature of political horse trading during your years in industry?

Allen Clamen: Tell me a little bit more about to what you're referring. The work that I have described was individual businesses looking at their portfolios from a business standpoint in order to ascertain how all the resources internally are being allocated. I'm not talking about, in this case, supporting a central corporate laboratory. That may be behind some of this.

Michael Schrage: Absolutely, but in both of those contexts the convergence of allocation of overhead is present. Given a portfolio and a particular perception, we want to charge a certain amount to overhead rather than another amount. If the portfolio is put in the incremental innovation category, we do a different overhead formulation than if it is pioneering research.

I've seen organizations blow themselves up over the political aspects of that debate. People from various parts of the organization emphasize things that are reflective of their background. I would like to see some sort of effort made to bridge a rational approach with what actually happens in organizations, since that ultimately is the theme that we're facing.

Allen Clamen: I may be naive, but I don't see that kind of infighting. Perhaps I had not seen it to the extent that you just described. I do know, however, that all costs are rolled in, so that the cost of resource development—an engineer, for example—might be significantly higher than his or her salary. I recognize that.

That's just recognition that resources cost money. We do have to house them since they do use a laboratory and supplies. That's all factored in to the overall budget. Within that budget framework, the projects are allocated regardless of the politics. Development resources cost this much, period.

Michael Schrage: And as far as you're concerned, how the overhead is allocated is a purely rational process.

Allen Clamen: For example, there is no allocation of marketing and administrative cost. We learned a long time ago that this is something we didn't want to get in to. There is a base cost of running the business that doesn't enter in to the cost of developing new products.

Kenneth A. Pickar, California Institute of Technology: One of the realities of managing portfolios is the immortal project that is almost impossible to kill. You think you've terminated it, and then you find out that it popped up in some other guise somewhere else in practice on the side. Sometimes these projects turn out to be great discoveries, but far more often they are really a terrible drain on resources. What type of special advice or ideas do you have for killing the immortal?

Allen Clamen: First, our gate decisions are taken at appropriate stages of activity or progress on a project to discern whether or not to continue. These stages are highly visible. Once the project is terminated, the people who are working on it will stop because the staff time and associated costs are dead.

Additionally, there are some companies we've benchmarked that actually celebrate a kill. It's hard to imagine. Everybody has worked on this. They spent the better part of their last year or two working on this. Now it's dead, and they're feeling like they've lost a best friend. Yet they recognize that the project is still in their knowledge base. There is now a jumping-off point into more promising things.

There are a number of ways to kill projects, but visibility is probably the best methodology.

Kenneth A. Pickar: The visibility is a function of how much money is spent. The projects can't pass the stage gate, but then they go back into the database. It is really an act of hibernation.

Allen Clamen: If you recall the New Concept Development Model, unused ideas return to an idea bank. There is an idea manager who is managing the bulk of ideas that are in there. If one of them does look like a recycled idea, that's fine.

Kenneth A. Pickar: They'll wait until you retire and then they'll propose it again.

Allen Clamen: Right.

Mary L. Mandich, Lucent Technologies: Do you think that portfolio management prevents that or that recycled ideas are just a natural human function?

Participant: I think portfolio management helps limit the amount of resources wasted on these kinds of issues.

Allen Clamen: That is part of the people issues.

Kimberly W. Thomas, Los Alamos National Laboratory: You said one of the keys to success is an environment that is very open to risk taking. How do you reward and encourage the risk taking for your individuals? I think you gave us just one idea: celebrating the death of a project.

Allen Clamen: ExxonMobil does not practice that, but I think it's a great idea. The other is that risk takers, if they've made a name for themselves, tend to be those who have had some success in the past. Recognize that one success takes a lot of failures. But this risk taker, whoever he or she is, has had a number of minisuccesses along the way. It's worth taking another chance with them. Motivation is very important. A company should not reward an employee simply because that person was part of a great innovation. An employee should be rewarded for good attempts. It is important to recognize good tries as well as successes along the way.

Additionally, companies need to say the organization encourages and supports risk taking. The industry has done the reverse for many years. We were averse to risk, which definitely discourages innovation. Just recognizing that a company must not be risk adverse to be successful is an important element.

David J. Soderberg, BP Chemicals: You mentioned a very coherent and concise set of tools that are used: portfolio management, fuzzy front end, and stage gate. Those work very well within a business unit context. They're very focused and allow you to allocate resources.

How do you get the cross-fertilization and the opportunities that are identified, in our terminology, between the different streams, namely upstream and downstream? In a broader context, how do you obtain cross-fertilization within our industry and outside our industry?

Allen Clamen: That's still a big challenge for us. As I mentioned, the portfolio management was all within the business context. There is a roll-up intended, which hasn't happened yet, among all businesses so that we can start to look across businesses. What does the whole profile look like? What does the company portfolio look like?

More importantly, we have a connection with our corporate research labs, which is a central organization where we try to share any innovation or technology that's promising or has been found to be useful in one business with the others. They become very good ombudsmen across businesses. We are fortunate to have that. In fact, it spans downstream and upstream because corporate has all those aspects.

Venkat Venkatasubramanian, Purdue University: The issues you raised today reminded me quite a bit about the portfolio management issues in the pharmaceutical industry. We work closely with Lilly and some others back at Purdue. One of the key items that continues to come up in the early stages of the drug development process is this notion of failing fast. They know that most of the ideas they pursue will fail. They believe that only one or two will actually make it all the way through the pipeline, which right now takes about 8 to 12 years. The company wants to identify those ideas that will succeed and those that will fail very quickly, and they call that "failing fast." They do celebrate killing those projects despite emotional ties. Is this notion strongly pursued in the nonpharmaceutical parts of product discovery?

Allen Clamen: Yes. I mentioned earlier that Stage 3 development activities take the bulk of the cost, time, and resource utilization. If a project can fail prior to that full development stage, you will be much more successful than if you had to waste resources to get to that point.

We give a great deal of effort to learning in the early stages. I called it up-front planning about the manufacturability of the products, the marketability of the products, and the home that these products would ultimately have in terms of our competitive advantage. This knowledge is important upfront rather than as you go through the process. The very early stages of basic research are a large investment. Recognizing that and making the decision early are absolutely critical. Failing fast, failing early—those are key concepts to making the whole enterprise more successful.

Michael Schrage: The term "failure" needs to be better defined. If the criteria for the project change in the course of research and development, the initiative hasn't failed; the hurdle rate has changed. It fails relative to malleable criteria, when in fact the underlying science may be quite valid and quite useful. Failure here is ironically too all encompassing, rather than focused. We need to see how our definitions of failure change.

The other thing is in terms of criteria. When we talk about risk, risk cannot be divorced from cost. When you have simulation tools, the cost of doing a test with computational chemistry is two orders of magnitude lower in the year 2002 than it was even 10 years ago. The idea that we talk about risk as some sort of fixed point is also dangerous. Additionally, the notion of our cost structures is also changing. Attention needs to be drawn to that explicitly.

Allen Clamen: What I would call a failure is the failure to recognize information that we had earlier that would have prevented the development that was wasted.

Mary L. Mandich: Could you comment on the human money resource versus the actual computers, workstations?

Allen Clamen: The ability we have in this day and age to use all the tools allows our people to do a lot more with less. That is critical.

3

The Chemistry Innovation Process: Breakthroughs for Electronics and Photonics

Elsa Reichmanis[1]
Bell Laboratories, Lucent Technologies

This chapter features the role of chemistry in the innovation of technologies that at first glance may not appear to be chemistry intensive, for example, electronics- and photonics-related advances. While the connection of these technologies that have revolutionized our way of life to disciplines such as electrical engineering, optical engineering, and computer science is readily discerned, materials chemistry also plays a significant role in their development and can be seen to provide an enabling foundation through materials and process design and development for desired functionalities.

Using the electronics industry as just one example, it is not an exaggeration to view semiconductor manufacturing facilities as large chemical factories. Chemistry has played a role in the scaling of silicon circuits for the past 50 years, from the invention of the transistor at Bell Labs (a device measured in inches) to current devices found in computers that have transistors with features as small as 130 nanometers. To put the size scale of current device features into perspective, an *E. coli* bacterium cell is about 1 micron by 5 microns in size, while a human hair is approximately 100 microns in diameter. In addition to decreased feature size, chemical processing has drastically reduced the cost of fabricating integrated circuits, thereby facilitating the computer revolution.

However, research and development related to electronics and photonics requires multidisciplinary approaches in order to ensure the development of materials and processes that are compatible with overall system needs. The enabling chemical advances that have occurred in these areas have not been made in isolation. Clearly the process of invention or knowledge creation is not one that is readily amenable to scheduling. It is simply not reasonable to expect "invention on demand" or for creation of fundamentally new insights or phenomena to take place on "schedule." On the other hand, innovation that uses existing knowledge and inventions to create new technologies can be facilitated. The most successful facilitation occurs through creation of an environment that encourages interactions among colleagues. In today's environment it is increasingly unlikely that a single individual will have the

---

[1] Elsa Reichmanis is director of the materials research department at Bell Laboratories, Lucent Technologies, in Murray Hill, NJ. She has also been elected president of the American Chemical Society, the world's largest scientific society, for the term beginning January 2003.

necessary breadth of expertise to turn an innovative idea into market reality. Multidisciplinary teams need to work in concert such that the real-time exchange of ideas, issues, and solutions results in concurrent development of multifaceted technologies. Their success depends largely on the commitment of the individuals to the overall program.

One of the key facilitators of advancements in microelectronics technologies was the formation of the Semiconductor Manufacturing Technology group (SEMATECH) in the 1980s. This organization provided a noncompetitive forum for discussion and identification of future trends and industry needs. More importantly, it served to establish research goals for this critically important industry sector and provided funding for precompetitive research in relevant areas. The Semiconductor Industry Association (SIA) Roadmap (see Figure 3.1) currently defines the proposed transistor feature size and related technology needs out to the year 2014. Additionally, it identifies roadblocks requiring research investment, thus fostering the necessary research and development to achieve manufacturable solutions.

Materials scientists and chemists alone cannot direct new technologies to commercially viable applications. Process and hardware engineers, tool manufacturers, and device designers need to be an

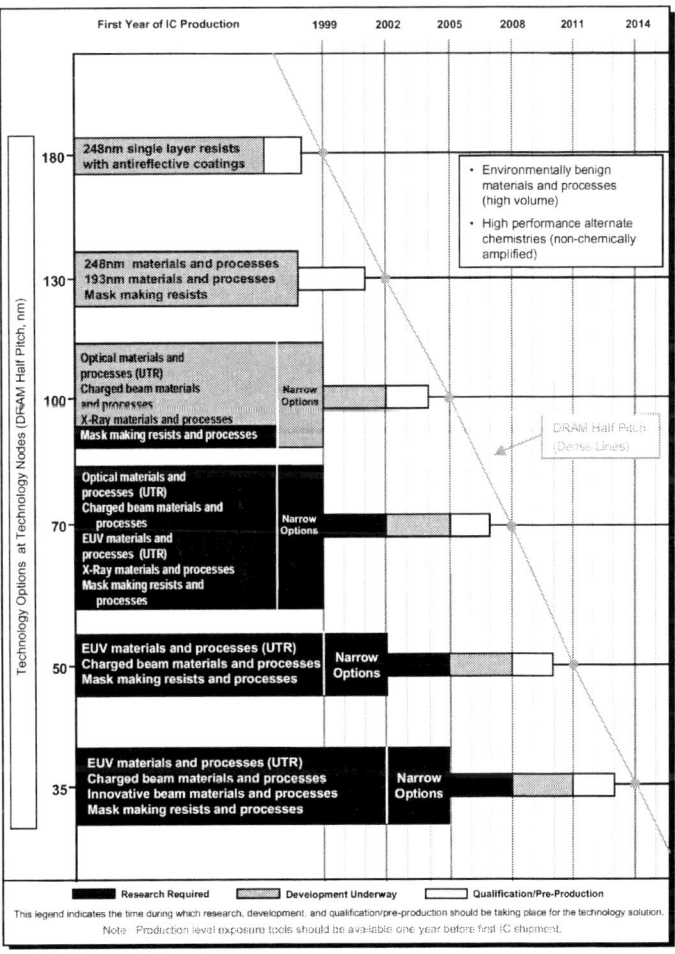

FIGURE 3.1 The SIA Roadmap provides direction to corporate semiconductor research laboratories (Semiconductor Industry Association, *The International Technology Roadmap for Semiconductors*, 1999 ed., International SEMATECH, Austin, TX.)

integral part of the team developing new materials for electronic applications. Recent advances in microlithographic materials design and development serve to illustrate key factors required for successful multidisciplinary technology development. The overall performance of resists used for lithography is based on the radiation response, resolution, linewidth control, defect density, etching resistance, adhesion, supply and quality assurance, shelf life, cost, and other factors. Through understanding of these requirements from a device manufacturing perspective, each of these parameters can be translated into specific molecular characteristics. For example, reducing olefinic and aromatic moieties for a material that needs to be transparent at ultraviolet (UV) wavelengths prevents the film from being opaque to incident irradiation. In the 1980s, Ito, Willson, and Frechet conceived the idea of using catalytic processes to overcome limitations in available light flux at the wafer plane during resist exposure.

The performance of the resist is controlled through the chemistries associated with each component. Using one 193-nanometer resist concept as an example, each component—the matrix resin, the dissolution inhibitor, and the photoacid generator—must be designed to be compatible with each other, but equally importantly, they must be compatible with the overall device fabrication process. Table 3.1 lists a number of materials requirements and the associated desired molecular characteristics.

In this example the matrix resin uses polymers containing alicyclic moieties, maleic anhydride, and acrylate derivatives. The alicyclic units provide for transparency and etching resistance, while acrylic acid and its derivatives can affect differential solubility. Maleic anhydride is incorporated into the polymer because it facilitates metal-ion free synthesis; metal contamination, even at minute concentrations, can degrade device performance. The dissolution inhibitor is a cholic acid derivative; it is a UV transparent, readily available steroid that occupies a large volume fraction of material, thereby enhancing the differential solubility in aqueous base of exposed and unexposed areas of the resist. The photoacid generator is a triflate diaryliodonium salt that is miscible with the resist components and generates a strong acid upon exposure to UV light while producing relatively nonvolatile byproducts upon irradiation so that the optics in the tool are not damaged. Each component was designed with broad understanding of manufacturing process requirements and hardware and device constraints: a necessary requirement for avoiding pursuit of materials chemistry concepts that would ultimately prove not to be manufacturable.

TABLE 3.1 Key Resist Materials Properties Related to Molecular Characteristics

Resist Property	Molecular Characteristic
Absorption	No olefinic or aromatic moiety
Etching stability	High level of structural carbon and low oxygen content
Aqueous base solubility	Presence of base solubilizing groups
Substrate adhesion	Presence of polar moieties
Sensitivity (photospeed)	Catalytic chain length for acidolysis, efficiency of acid generation, acid strength, protective group chemistry
Process latitude and substrate sensitivity	Catalytic chain length for acidolysis, acid strength, protective group chemistry
Outgassing	Protective group and acid generator chemistry
Aspect ratio of images	Surface tension effects and mechanical strength of materials
Low metal ion content	Synthesis and scale-up methodology
Manufacturability and cost	Synthesis and materials scale-up methodology, lithographic process requirements

The concept of using UV wavelengths for lithography (deep-UV lithography) was introduced in the mid-1970s with the demonstration that poly(methyl methacrylate) could be imaged upon UV irradiation. Furthermore, Bowden and Chandross demonstrated imaging of poly(1-butene sulfone) upon exposure to 185-nanometer light.[2] Although deep-UV lithography was not practical at this point, the work led to interest in the design and development of short-wavelength sensitive materials for lithographic applications. Ultimately, in the early 1980s, chemically amplified resist strategies were introduced, leading to the commercial introduction of 248-nanometer resists in the early 1990s. As interest in using still shorter wavelengths for lithographic images grew, research associated with the design of materials chemistries for 193-nanometer exposure was initiated. While the introduction of 193-nanometer resists was made possible by the knowledge gained in the early 1980s, serious work in materials development began in the early 1990s. Although attempts were made to shorten the development cycle, the time cycle for market introduction of a materials technology remained at roughly 10 years from initiation of direct applications-focused research. Resists using lasers at still shorter wavelengths will have even larger challenges associated with the resist materials properties, and it is likely that future timescales covering exploratory research, applications-oriented research, applied development, and market introduction will remain in the range of a decade.

Looking at another materials-intensive technology (one that is yet to be commercialized), similar trends surrounding multidisciplinary team interactions and time lines emerge. Conducting polymers have long been of interest for both the fundamental research community and the commercial sector. The first demonstration of semiconducting behavior in plastics took place in the late 1980s in Garnier's laboratory in France. These exciting results led to a surge of interest in such materials fueled by the prospect of fabricating thin, flexible, lightweight devices using low-cost printing techniques. In 1997 the first printed all-plastic transistor was demonstrated,[3] and in 2000 the first large-scale complementary circuit with 864 transistors was fabricated.[4]

More than a decade after Garnier's initial discovery, a team of researchers from Bell Labs and E Ink Corporation envisioned a prototype plastic display that could be considered the first demonstration of electronic paper. This program was initiated in late 2000, and within a year the first "plastic paper" was demonstrated using a 256-transistor back plane fabricated on a sheet of mylar that was then laminated onto a similar sheet of an electrophoretic display material.[5] This achievement was the result of a technology-focused, multidisciplinary team of scientists and engineers working together from the outset to identify not only the end goal but also each step required to reach that goal. The need to understand the interplay between device parameters, materials performance, and process technologies was implicit. Further technology-focused research will be required before "plastic electronics" will reach the commercial sector. To be successful, a cohesive multi-disciplinary team is a requirement.

Our communications infrastructure relies heavily on advanced materials chemistries. From the manufacturing processes used to fabricate optical fiber cables to molecular beam epitaxy techniques for the creation of nanoscale heterostructures that enable many optical devices, innovations in materials chemistry have played a role. An example of a recent technological achievement that relates to optical communications systems is the MEMS-based (microelectromechanical system) Lambda Router. The Lambda Router is an optical system developed at Lucent Technologies for switching narrowly focused

---

[2]M. J. Bowden and E. A. Chandross. 1975. Poly(vinyl arene sulfones) as novel positive photoresists. *Journal of the Electrochemical Society* 122: 1370-1374.
[3]<http://www.bell-labs.com/org/physicalsciences/timeline/1998_transistor_expansion.html>
[4]<http://www.bell-labs.com/news/2000/march/20/1.html>
[5]<http://www.lucent.com/press/1100/001120.coa.html>

beams of light in the core of an optical network. The idea for using micromachines in light-wave networks was conceived around 1990, and the first example was a mechanical device that deflected light. Over time, technology concepts matured and evolved into the present system that uses complex arrays of small MEMS mirrors. In a fully connected system, light can be directed from one point to any desired point in the optical system.

The connection to chemistry comes about through the materials, process, and packaging technologies used to fabricate the devices. Microelectromechanical devices are fabricated using silicon-based processing, and thus the processing, packaging, reliability, and manufacturability all depend on a chemical knowledge base. Integrated cross-disciplinary teamwork is essential to ensure understanding of all relevant systems parameters and to minimize costly late-stage design changes while helping to focus the system and chip designs toward a viable commercial product.

The implementation time line for MEMS-based mirrors is very similar to those discussed above. Using the Texas Instruments planar digital mirror device as an example, the development to product implementation cycle time was 5 years, preceded by 10 years of active technology-focused research, not to mention the very relevant silicon-based process research that preceded the development of MEMS technologies. Once again, elapsed time from research concept to market implementation was in the range of 10 to 15 years. Notably, the time frame from concept to market implementation for the Lambda Router using a three-dimensional MEMS mirror design took only 18 months. In this case the team had a clear goal with defined technology milestones and was well coupled to a customer through marketing and sales professionals who were integrated into the team. However, the aggressive development cycle for the Lambda Router would not have been possible without preceding enabling research and development efforts related to MEMS mirror arrays and other optical devices. Even these efforts were backed by the demonstration of microgears in the 1990s based on MEMS technology, which was additionally supported by 30 years of integrated circuit research and development experience.

Using the three examples described above as references, there is no question that materials chemistry plays a critical role in the development of advanced technologies. Each case has relied heavily on both fundamental and long-range technology-focused research for enabling ideas, concepts, and inventions. While it is not reasonable to expect that the creation of fundamental insights can be dictated by a predetermined schedule, once relevant inventions and enabling technological concepts are realized, the innovation process can be facilitated through the use of integrated multidisciplinary teams to rapidly capitalize on research insights. These teams need to communicate effectively, understand the value that each individual brings, trust each other, and, perhaps most importantly, understand market needs. In a similar vein, marketing, sales, development, and manufacturing organizations need to have knowledge of the latest scientific findings as well. Decreasing the time line for innovative new technologies requires more than just carrying out the requisite materials chemistry research and development; it involves identifying and strengthening the connections among all the stakeholders. The end product as well as the steps to achieving that end product must be defined with a view toward manufacturability.

DISCUSSION

Mary L. Mandich, Lucent Technologies: First, who started the Semiconductor Industry Association (SIA)? Also, is there an equivalent organization in the more mainstream chemical industry?

Elsa Reichmanis: I can't directly speak to the origins of SIA, but SEMATECH was formed in the mid-1980s in an effort to regain U.S. leadership in semiconductor technology. Funding was provided by the Department of Defense and leading semiconductor industries.

SEMATECH was started in direct response to a feeling that silicon-based technology was being taken out of the country, that Japan was largely dominating the industry. We wanted to be sure that we retained a healthy industry in the United States. Interestingly, SEMATECH has become more of an international organization as companies have become more global in nature.

I am not aware of a comparable activity in the chemical industry. There is a battery-oriented consortium that the automotive industry belongs to. I know there is a similar packaging consortium, but this is again more silicon technology based. I think that because the chemical industry is so diverse, any chemical consortium would have to be technology oriented rather than broadly chemical in scope.

Hans Thomann, ExxonMobil: Historically, Bell Labs is known for championing the individual as a team. I am wondering, in the context of the examples you gave such as plastic electronics, if there were many years of research during which ideas were created for applications of that research. Are you using innovation processes to facilitate a conversion from this past mode of research guidance to the present market-driven research, and what tools are you using to help your multifunctional teams reach those goals?

Elsa Reichmanis: That's an interesting question. I can't say that the research management at Bell Labs uses a defined process for encouraging innovation. I think that we have a dual personality. We have a very strong individual contributor component where individuals are rewarded for their accomplishments. At the same time, many of those individuals traditionally have—and I would say more so today—also participated in team-based activities. If you look at the publications coming out of Bell Labs, I think you're going to be hard pressed to find many papers with a single author.

While we are awarding the individual, at the same time we have a very collaborative environment that encourages team approaches to understanding and solving problems. We also have an open-door policy. We have wide corridors that are very long, and most people leave their office doors open. You can easily talk to your neighbor. You can find somebody with the right expertise down the hall or in another building to talk to about a problem that you can't solve yourself. The culture is one that encourages the interaction of communication.

Hans Thomann: Are you using any formal processes for innovation, such as targeted innovation sessions?

Elsa Reichmanis: We don't use formal innovation processes such as those described in the preceding presentations. We do continually examine our portfolio of activities and determine what relates to the business from the advanced development, applied research perspective and from the very long-term research perspective. We need to have a mix of both, and we have a broad continuum of activity. In reality the spectrum of fundamental research to productization requires different modes of working at different stages—there is an evolution from what could perhaps be more individual blue-sky research to a team-based, problem-solving approach. To be successful, we need flexibility.

I'm in a laboratory that has a mission for doing research that should produce applications 5 to 15 years in the future. Company executives accept that long-term research in the advanced technology arena is necessary for the company to continue to be a provider of advanced technologies in the future. Otherwise, we won't have a business.

Even in a very tough economic climate for Lucent, there is resounding support for long-range fundamental research that does not necessarily have a direct application to the business. The research will generate knowledge, and that knowledge will generate ideas for technology advancement.

David E. Nikles, University of Alabama: SEMATECH produces road maps that identify the key problems to be solved within approximately 10 years. Is that why the corporate research timescale seems to be 10 years?

Elsa Reichmanis: I don't believe so. Historically, the timescale has always been 10 to 15 years from research concept to product implementation. I don't think the 10-year SEMATECH road map is defining this cycle.

David E. Nikles: We've heard about the innovation cycle from three big companies that have established businesses. How long was that from discovery to marketplace for the polymerase chain reaction for replicating DNA? Do smaller companies have a faster pace?

Elsa Reichmanis: I have read a lot about start-ups and smaller companies having a faster pace and being the "innovators of the future." I don't entirely buy into that. I think you can have that same rapid time to market if you have the right environment in a large company. For example, Lucent's MEMS Lambda Router took only 18 months from idea to delivery of product, and I don't think of Lucent as a small company.

If we start relying on small companies to produce advanced technologies, the industry will be hurting 10 or 15 years down the road. Small companies are very focused on the development end of technology and ensuring that they have a revenue stream. They don't have the resources to do research in broad areas. Companies need to have a balanced portfolio that includes a number of different avenues for doing fundamental research in and of itself; fundamental research as it may apply to a technology, industry, and applications; and applied research and advanced development activities. We need to have a balanced perspective of the research path.

Robert A. Beyerlein, National Institute of Standards and Technology: My question is very close to Hans Thomann's. I'd like for you to comment further on what you think are the essential elements at or near the beginning of the innovation process.

I would like to mention that in Allen Bard's Priestly Award address he expressed the opinion that new science is originated from one or two investigators working largely in isolation. At the same time, he made comments showing his skepticism about the trend he saw toward funding collaborative centers by the federal government and less emphasis on supporting the individual investigator.

You gave three powerful examples of the route from innovation to effective implementation in the marketplace and how, early on, teams were and are important. In the spirit of Allen Bard, could you comment further on what the essential elements at or near the beginning of innovation are and how these have changed?

Elsa Reichmanis: First we have to differentiate innovation from invention and discovery. I agree with Allen to some extent that invention and discovery can be done either by an individual scientist working in isolation or by groups of scientists working together.

The electronic paper idea and the work we've done in organic semiconductors have had components including work by individual scientists, but there have also been components of a multidisciplinary team approach. For example, chemists have been associated with the design and characterization of the semiconducting organic materials, developing the understanding of what functionalities are needed and other factors.

The chemists have worked very closely with the physicists and the device engineers interested in understanding how these devices work. In turn, they have all worked with process engineers to define reproducible fabrication processes. Each question, such as "How can a device be designed for better performance?" or "How can the material be processed in a more cost-effective manner?," is answered by an individual inventor. However, all the scientists come together as a team to provide a technology objective. I don't completely agree with what Allen Bard said, and I don't disagree either. I think there needs to be a balance between individual and teamwork. I believe the academic sector should place more emphasis on teaching how to successfully work in teams.

Many exciting new discoveries will be at the interfaces between disciplines. To work at the interface we need to be able to facilitate interactions and understand how to talk to our colleagues. Physicists, biologists, and chemists don't necessarily talk the same language. We need to learn how to communicate with each other effectively in order to be able to drive innovation in the future.

Lawrence H. Dubois, SRI International: As an alumni of Bell Labs and a current resident of Silicon Valley, I can tell you that the SIA Roadmap is an incredibly powerful tool. It drives a number of different industries to focus on where they want to go and a number of different disciplines to develop new kinds of resists, etching tools, lasers, deposition chemistries, and the like. Having said that, the SIA Roadmap also stifles creativity. If you come up with an idea or a concept 5 years too early, nobody cares. It will sit on the shelf until it's time for it to be on the road map. There are clearly pluses and minuses to road maps.

Elsa Reichmanis: That's happened to Bell Labs a lot, too. We've come up with inventions too early.

Kenneth A. Pickar, California Institute of Technology: Let me just throw another stone on this one, too. Gordon Moore, who predicted that data density would double every 18 months, would be the first to tell you it was not a stroke of genius on his part. Things like the road map are a self-fulfilling prophecy. Creativity may have been stifled, but maybe if you look at how the business has expanded, it's hard to see how it could have been done any better.

Your talk is astonishing to me. I read the financial section, and your business is in meltdown. It is the worst catastrophe in the history of the telephone since Alexander Graham Bell. To hear you talk, Elsa, it's like business as usual in the research laboratory. That's just amazing. Who do you talk to out in the businesses when they're out there firing people?

Elsa Reichmanis: Life has certainly changed, but Bell Labs really benefits from the executive level's belief—meaning Patricia Russo, CEO, Lucent Technologies; Bill O'Shea, president of Bell Labs; Jeff Jaffe, president of research and advanced technologies for Lucent; and the board—in supporting long-range, fundamental research.

This belief is demonstrated by their commitment of dollars to the research organization. We have laboratories with a shorter-term focus that are more aligned with interactions with the business unit, meeting not next-generation but perhaps second-generation, third-generation needs.

Mary Mandich is involved in one of those efforts, where the bulk of her work, but not all of it, involves a shorter-term, applications-driven research program. It's not a development program but is somewhere between applied research and applications-driven research. There are also some fundamental research aspects along the way that could enable implementation of a new technology.

On the other hand, the Physical Research Lab that I work in is looking at the longer term. We are very well funded, particularly in light of the company's economic situation. We don't have enough money to do all we'd like, and we are a much smaller organization than we were 5 or 10 years ago. We have to worry about whether we can maintain adequate critical mass to do what we want to do and to provide technology to the company 5 or 10 years from now. The company appreciates our research efforts and continues to support us.

I was at a Kellogg technology management program a couple of years ago, and one of the presenters said there are two kinds of chief executive officers: those who don't really understand technology and therefore don't believe in it and are very critical of research activities and those who take technology as a matter of faith. For a technology business to exist 10 or 20 years from now, the business must have a research program. Otherwise, new technologies will not be available to the company and it won't be competitive and stay in business in the long term.

Believing in, valuing, and needing technology is a matter of faith.

4

DARPA's Approach to Innovation and Its Reflection in Industry

Lawrence H. Dubois[1]
SRI International

INTRODUCTION

Today's world is changing rapidly, providing exceptional challenges and opportunities. As shown by recent events, it is increasingly complex and chaotic with seemingly small actions triggering massive changes. In addition, the rate of technological change is accelerating at what some would say is an exponential pace based on principles such as Moore's law, Metcalf's law, and Schumpeter's waves. An outcome of this change is that our work is becoming more interdisciplinary. Information technology is impacting chemistry, physics is impacting biology, and nanotechnology is pervasive in many disciplines.

How does one manage and control these changes? How does one harness this complexity and growing multidisciplinarity to solve critical problems for society? For approximately 50 years the Defense Advanced Research Projects Agency (DARPA) has played a leading role in turning innovations in technology into new military capabilities. In fact, most military and many civilian systems today can trace their origins to funding from DARPA. These include the Internet (ARPANET), high-speed microelectronics, stealth and satellite technologies, unmanned vehicles, and a wide variety of new materials. What are the driving forces, culture, and processes employed by DARPA to accelerate technical innovation, and how can these same techniques be used effectively in academia, national laboratories, and industry?

DARPA

In order to understand DARPA, one must understand the context in which it operates. DARPA is the central research and development organization of the Department of Defense (DOD) and has a very

---

[1] Lawrence Dubois joined SRI International as vice president and head of the Physical Sciences Division in March 2000. Prior to that, he spent 7 years at DARPA, finishing his tenure there as director of the Defense Sciences Office, which is responsible for an annual investment of approximately $300 million toward the development of technologies for biological warfare defense, biology, defense applications of advanced mathematics, and materials and devices for new military capabilities.

specific mission: innovation in the defense of our country. Other agencies and offices in the defense and intelligence communities play different and complementary roles. The same is found in all large institutions. Peter Drucker[2] explained that there are three types of changes that occur in all complex organizations. The first is systematic continuing improvement, which the DOD calls training and experimentation. The second is based on building tomorrow's systems using today's proven techniques and technologies. The DOD views this as evolutionary requirements-based research and development as practiced by, for example, the Office of Naval Research and the Army and Air Force Research Laboratories. The third type of innovation required by any healthy organization is innovation with a goal that makes obsolete and to a large extent replaces even the most successful current products and processes. For the DOD the function of *radical innovation* is carried out by DARPA.

Innovation is more than invention—it is invention turned into practice, and it requires a fundamental change in operations. Innovation is a very slow process in most organizations, and it is especially slow in large institutions where continuing success can breed a risk-averse atmosphere. Furthermore, *radical innovation* is risky and requires real leadership, dedication, and protection from above. How does one do this consistently in an organization like DARPA that invests over $2 billion per year in advanced research and development?

Despite its ties to the Pentagon, DARPA's strategy is to remain small and flexible and to quickly exploit emerging technologies and situations. DARPA has a broader horizon than most commercial analogs, such as working with venture capital firms, but it is more focused than traditional university-based research. DARPA is not bound by military *requirements* (official military doctrine) but rather responds to military *needs*. For the most part, DARPA emphasizes high technical payoffs for which success may provide dramatic advances in military capabilities. This usually entails taking high risks and focusing investments in a few critical areas. In this sense, DARPA is more like an investment firm since it has no long-term investments in "bricks and mortar" (no in-house research labs) and no established constituency that it must "keep fed."

DARPA is a small, relatively flat organization with approximately 120 technical staff, 220 total employees, and only one level of management between the program managers and the director of the agency. This allows ideas to flow very quickly. Projects, program managers, and even the agency director rarely last more than 3 to 5 years, and there are seldom renewals. This constant flux of programs, program managers, and directors leads to a rapid generation of new ideas. Because of limited resources, there is competition for funding the best new ideas, both internally and externally. Each project is managed by a proactive program manager, and quality performance is rewarded with increased funding. In order to accomplish this, DARPA has highly flexible contracting and hiring practices that are atypical of most of the federal government. DARPA contract agents can issue contracts, grants, and various other transactions. Staff can be hired from industry quickly, at wages substantially above those of typical government employees.[3]

---

[2]Peter F. Drucker. 1998. Management's new paradigms. *Forbes* (October 5):152-177.

[3]DARPA has at least two ways to bring new staff in from industry in addition to hiring as government employees. The first is the traditional Intergovernmental Personnel Act route. The initial step in this process is to associate the candidate with an eligible institution (e.g., college or university, federally funded research and development center, not-for-profit, etc.). According to the act, DARPA may hire qualified personnel from these organizations for a limited period without loss of employee rights and benefits. Appointments are generally from one to a maximum of 4 years. This process works and is how the author was initially hired at DARPA.

More recently, DARPA has been granted Experimental Personnel Hiring Authority under Section 1101 of the Strom Thurmond National Defense Authorization Act for Fiscal Year 1999. Under this authority, DARPA can directly hire up to 40 eminent scientists and engineers from outside government service for term appointments up to 4 years. These appointments may be extended to 6 years in specific cases. This authority significantly streamlines and accelerates the hiring process.

DARPA has a set of very strict investment criteria. There are seven key questions that must be answered by each program manager and that in turn must be answered by individual project leaders or principal investigators:

- What are you trying to accomplish?
- How is it done today and what are the limitations?
  - What is truly new in your approach that will remove current limitations and improve performance? By how much? A factor of 10? 100? More?
- If successful, what difference will it make and to whom?
  - What are the midterm exams, final exams, or full-scale applications required to prove your hypothesis? When will they be done?
  - What is the DARPA "exit strategy?" Who will take the technologies that you have developed and turn them into a new capability or a real product?
- How much will it cost?

How do DARPA program managers differ from those in other funding agencies, and how do their efforts reduce the time it takes to go from basic research to innovation? First, the role of a DARPA program manager is different than that of most of his or her colleagues in larger, more traditional government funding organizations. In the Defense Sciences Office of DARPA, program managers must be proactive "techno-scouts" constantly searching for the next big technological opportunity. He or she is constantly talking to potential new contractors as well as possible users of any new capability. Once a new opportunity is identified, the goal is to grow this discovery with a judicious amount of money and technical talent and a modicum of oversight to catalyze the creation of a new capability. Since the tenure of a typical program manger at DARPA is on the order of 4 years, this must be done very quickly. Thus, efforts are highly focused, and goals and military needs are clearly understood by all up front. To accomplish this, DARPA program managers are given both the responsibility *and* the authority to act. There is both technical and fiscal flexibility, where the goal is to develop a new capability, not to fund someone's pet project for years.

The Defense Sciences Office is technically diverse and highly interactive, which naturally leads to collaboration and multidisciplinary projects. Many of the most interesting opportunities are at the interfaces between conventional disciplines. This working environment is conducive to such activities. Multidisciplinarity is also accomplished through the teaming of universities, service and federal laboratories, small businesses, large industry, etc. This allows one to develop a portfolio of technologies by combining basic and applied research with development and demonstration. By working synergistically with industry and pairing experts in fundamental research with those charged with producing a product, technology is transitioned more rapidly from the laboratory into the marketplace. As noted above, projects do not go on forever, and therefore DARPA program managers are always developing an appropriate exit strategy. Thus, DARPA technical staff work closely with business/industry leaders and department acquisition officials to ensure a market pull for the technology. Transitioning a research program into a long-term funding opportunity for the same group of contractors by another government agency is not the preferred exit strategy.

Program management at DARPA is a very proactive activity. It can be likened to playing a game of multidimensional chess. As a chess player, one always knows what the goal is, but there are many ways to reach checkmate. Like a program manager, a chess player starts off with many different pieces (independent research groups) in different geographic locations (squares on the board) and with different useful capabilities (fundamental and applied research or experiment and theory, for example). One uses this team to mount a coordinated attack (in one case to solve key technical problems and for another to defeat one's opponent). One of the challenges in both cases is that the target is continually moving. The DARPA program manager has to deal with both emerging technologies and constantly changing customer demand, whereas the chess player has to contend with his or her opponent's king and surrounding players always moving. Thus, both face changing obstacles and opportunities. The proactive player typically wins the chess game, and it is the proactive program manager who is usually most successful at DARPA.

PROGRAM MANAGEMENT DARPA STYLE

The traditional method of technology development where lengthy proposals are written and submitted to an august group of peers for review is incredibly time consuming and leads to a very inefficient use of resources. In this case, research by competing individuals working in isolation leads to a vast array of potential technologies and discoveries, only a fraction of which are ever combined to form useful new products and/or processes (see Figure 4.1). Most published papers sit idle in the archival literature with few, if any, references. Movement of technology out of the research lab and into the marketplace is generally slow. An example of this lengthy process is the development and application of new materials. Materials development is typically highly empirical and appropriate, and coordinated experiments and modeling are rarely done early on in the research process to answer critical questions an end user might have. The disconnect between researcher and application engineer is reflected in the amount of time needed between an idea and an end product. For example, to build a reliable part out of a known alloy requires a minimum of 36 months. This is short compared to the time it takes to change ship steel (7 to 10 years), apply lightweight composites (more than 15 years), or develop ceramics for engines (more than 20 years).

In order to circumvent some of the inefficiencies inherent in traditional funding organizations, the concept of a technology road map has been developed. Unfortunately, a road map is not always an appropriate solution. While road maps have several good points, including providing direction, defining

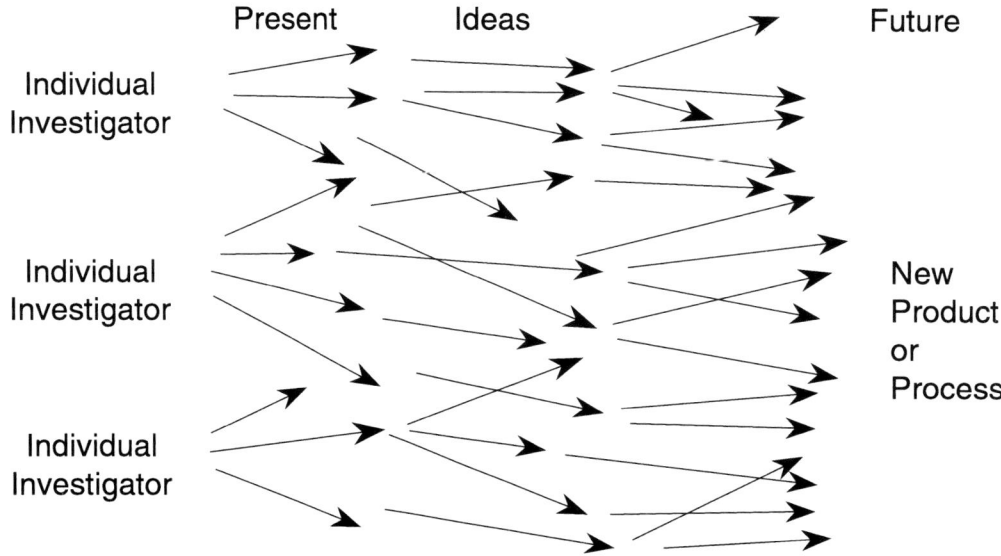

FIGURE 4.1 The traditional approach to technology development.

distances, and supplying a plan to get from point A to point B around obstacles, they do not provide a complete explanation of how research or even technology development should be done. For example, a road map assumes that everyone starts from the same place and that the destination remains fixed. In other words, there is no competing solution being developed and user needs do not change. It also assumes that no new roads will be built or that an airplane will be invented. Road maps can stifle creativity and do not account for serendipity.

An alternative to the standard method of performing research is termed the "end-game" approach and is typical of many DARPA-funded programs (see Figure 4.2). By first defining the desired product or process and the anticipated technology needs, research teams can better coordinate their efforts and a higher rate of return on technology development can be realized. The results of fundamental research are tied to the needs of the technologists, who then build on this information to further create new and useful knowledge. Basic research, applied research, and development and demonstration play a role at all levels in the process since there is a tight feedback loop between discovery, whether planned or serendipity, and end use. Frequent contact between technology developers and technology users, with the DARPA program manager playing the role of "technology midwife" at times, ensures that useful discoveries will move more rapidly from the research laboratory into the marketplace.

The development of a practical liquid feed for a direct methanol oxidation fuel cell provides a useful example. The reaction chemistry ($CH_3OH + 1.5O_2 \rightarrow CO_2 + 2H_2O$), and much of the basic technology has been known for decades. Despite substantial investments by both industry and the government, little progress was made. By using the end-game approach, DARPA program managers drove the process to rapid success—they obtained a several order-of-magnitude improvement in performance in a matter of a few years. Catalyst discovery and optimization, an understanding of fundamental electrochemical kinetics and modeling, and polymer membrane chemistry all played a key role at different stages of the process (see Figure 4.3). In contrast to the more traditional approach to technology development, it is the coupling of the research teams from academia, federal laboratories, and industry as well as across different disciplines that led to this rapid success.

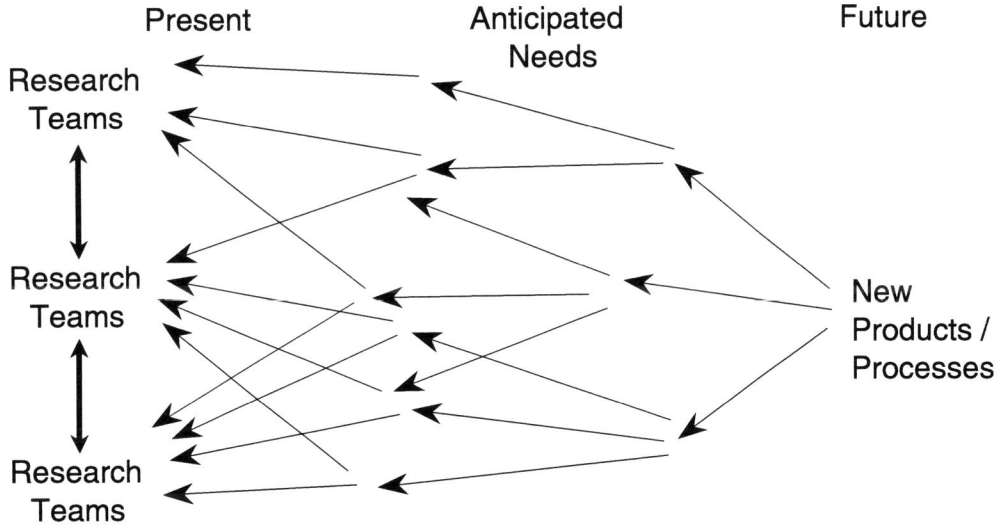

FIGURE 4.2 The "end-game" approach to technology development.

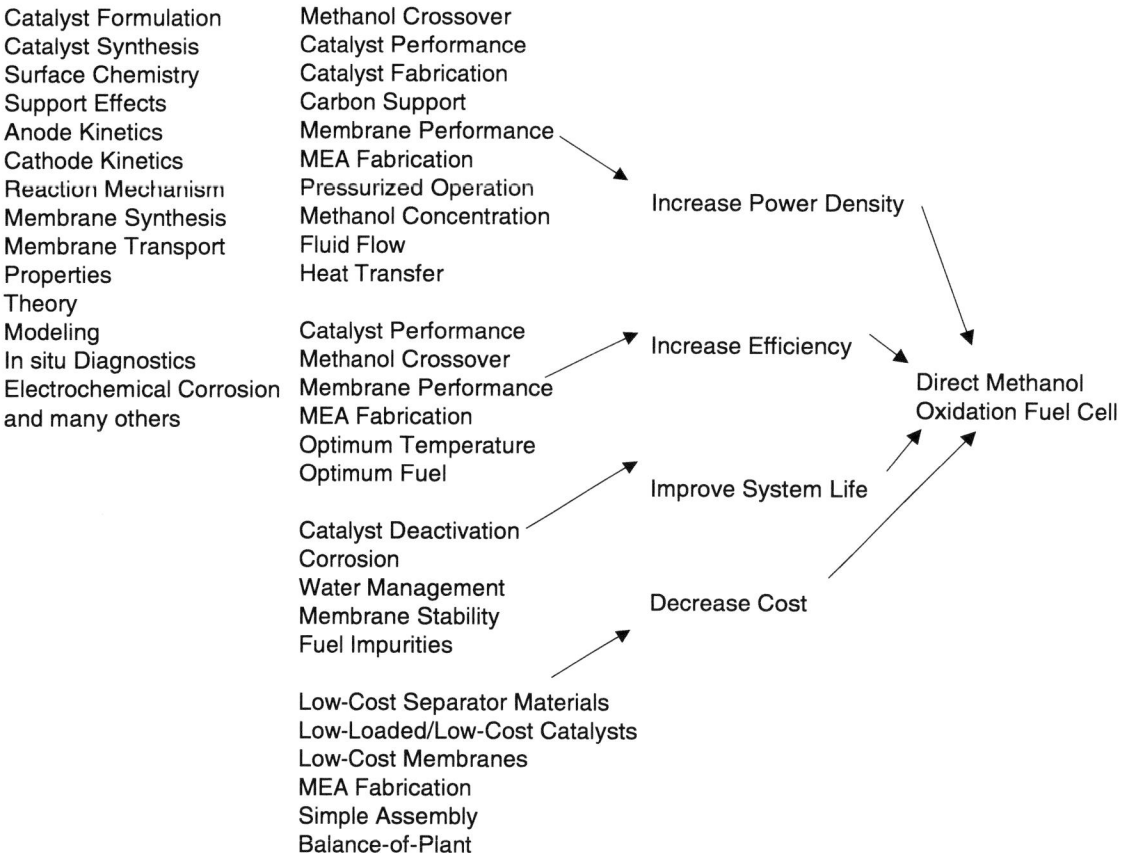

FIGURE 4.3 Research and development issues in the development of direct methanol oxidation fuel cells.

THE TECH TRANSFER PROCESS

Once a good idea—a true "golden nugget"—is found, effective technology transfer is a multidimensional process. Many of the key components for successful technology transfer are outlined in Figure 4.4. Some of these may be enhanced or driven by a funding organization, while others must be led by parties more closely involved with the technology. Just lobbing the technology "over the fence" and hoping someone picks it up rarely works. The key to success is knowing what to do. Should the project further enhance an existing technology with a well-defined market or should it commercialize a completely new product or process? Are there licensing possibilities into a well-established industry or is it necessary to build a large infrastructure in order to compete effectively? Is the starting point a profitable company with an existing manufacturing capability or the formation of a new company from the results of university- or national laboratory-based research?

The license versus venture decision is a critical one when trying to accelerate the movement of basic research into the marketplace. A number of important factors come into play. For example, licensing works best for the development of an improvement to an existing product or process. In this case, there are typically a handful of established players in the market and generally the barriers to entry are high. Alternatively, if development of a revolutionary technology creates a new complementary product opportunity, licensing may also be the most appropriate method of bringing an innovation to market. Frequent and early contact with the expected user is critical to moving technology to the market more

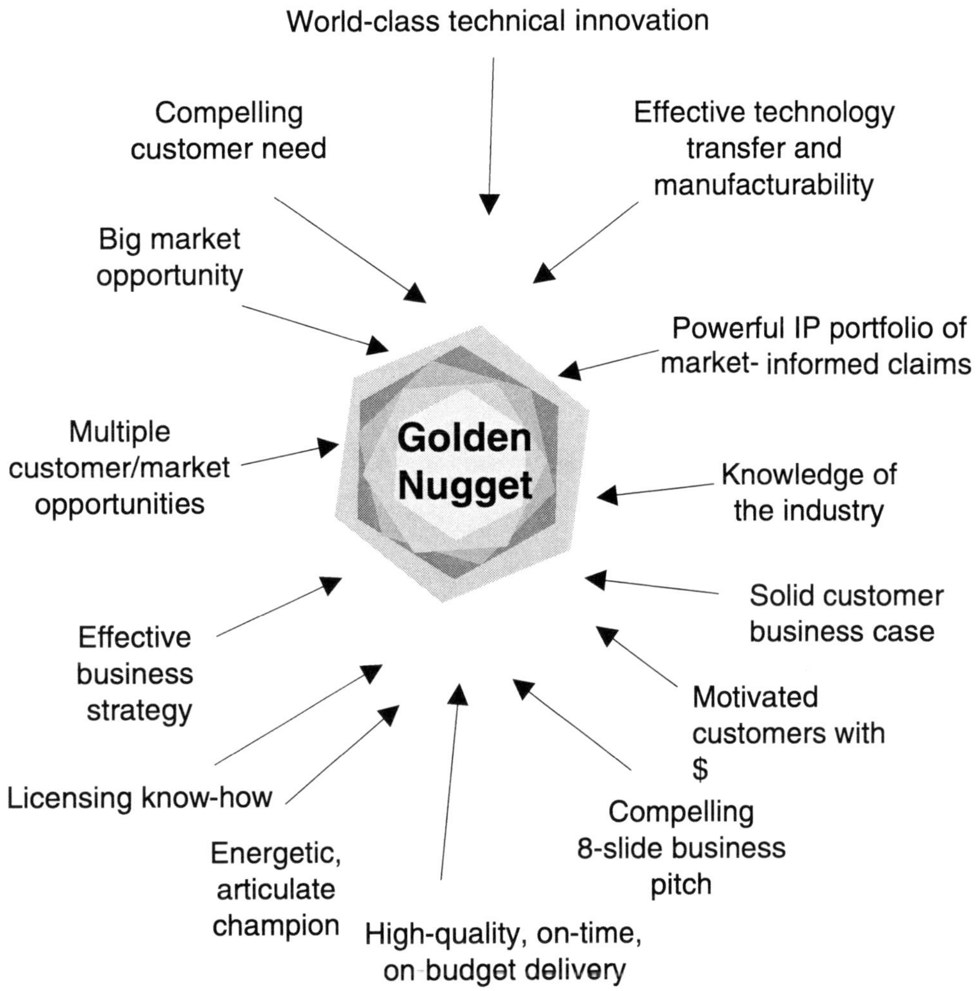

Diagram: "Golden Nugget" at center, with arrows pointing inward from the following surrounding elements:
- World-class technical innovation
- Compelling customer need
- Effective technology transfer and manufacturability
- Big market opportunity
- Powerful IP portfolio of market-informed claims
- Multiple customer/market opportunities
- Knowledge of the industry
- Solid customer business case
- Effective business strategy
- Motivated customers with $
- Licensing know-how
- Compelling 8-slide business pitch
- Energetic, articulate champion
- High-quality, on-time, on-budget delivery

FIGURE 4.4 Necessary aspects to enhance technology transfer once a potentially good idea ("golden nugget") is identified.

rapidly. The movement of people—for example, students in the case of university-based research—also speeds this process. It is said by many that effective technology transfer is a contact sport, and the more contact the better.

Starting a technology venture is more difficult and longer than licensing a new technology, but the payoff can be much greater. For most ventures the market needs must be sizeable, on the order of $1 billion. Venture capitalists typically want a very large opportunity, for which a company can be valued at greater than $100 million in less than 5 years and for which significant market share is possible. They expect a return on investment of 30 to 40 percent per year and breakeven in a reasonable time. The venture community is generally risk averse and is willing to take market risk but not technology risk. Thus, most venture-funded technologies are at a fairly mature stage. In addition to technology, a compelling competitive advantage and solid intellectual property protection are needed. Technical and business champions are a must, as is a dedicated team. Despite the typical work ethic at a venture-backed start-up, all this takes a substantial amount of time. Again, frequent feedback from the market is critical to accelerating success.

SRI INTERNATIONAL

In many respects, work at SRI International mirrors the way DARPA manages programs. SRI, one of the world's premier contract research and development organizations, has been delivering innovative science and technology solutions to governments and businesses worldwide for over 50 years. Like DARPA, SRI brings multidisciplinary teams consisting of technical depth spanning many fields, along with business and market insights to meet complex challenges. These teams are led by technical champions, individuals who have the passion to make something important happen and who work across traditional organizational and disciplinary boundaries. Since SRI uses its research for organizations like DARPA to develop commercial opportunities, it is focused on moving technology rapidly from the research laboratory into the marketplace. Staff at SRI endeavor to have all of the elements outlined in Figure 4.4 in place in order to reduce the time it takes to move basic research from the laboratory and turn it into a true innovation.

SRI has created a number of tools in order to help its staff speed the technology transfer process. These include the NABCs, a simple way to capture the true impact of an effort. N stands for the customer, client, or market *need*. This can be commercial, government, or societal. A is the compelling technical *approach*, which can be new science, new engineering, or new theory. B is the *benefits* that would accrue if one were successful. C is the worldwide *competition*, reminding the primary investigator or project manager to check who else is doing similar work so that other efforts are not duplicated. The key to success is not just developing a great technical approach but a thorough understanding of the market needs and the competition. This includes not only where the competition is today but also where it will be when the new product or process comes to market. Note the similarity between the NABC formalism used by SRI and the seven questions asked of all DARPA program managers.

SRI has developed a series of specialized "watering holes" or gathering places where staff can present and vet their ideas in an open, mutually supportive forum (NABC presentations). Typical watering holes span multiple disciplines and include business development staff in addition to scientists and engineers. As ideas mature, SRI has developed an online Business Development "Cookbook," a how-to guide to move technology into the marketplace and to build relationships with government and commercial clients. The SRI Internal Ventures and Licensing Board reviews, evaluates, and nurtures emerging business opportunities and provides a forum where business leaders can supply feedback on emerging technologies and markets. A formalized royalty and equity-sharing program that rewards staff for the value they create is an added incentive to help speed commercially successful products and processes into the market. Reinvesting the remainder of the funds received from licensing or equity in new equipment and facilities helps to keep the facilities at SRI state of the art. This helps to deliver more technology more quickly to the customers and to attract talented staff.

RECOMMENDATIONS

Blindly funding the technology transfer process is clearly not the most effective answer to improving the ability to move technology out of the laboratory and into the marketplace. Funding, which is always a limited resource, must be invested wisely. Based on my experience at Bell Laboratories, DARPA, and SRI International, I suggest three cross-cutting themes that affect all of the issues outlined above and seriously enhance or impede the speed with which technology commercialization can occur: (1) focus on important problems, (2) keep the end in mind, and (3) empower funding organizations. Since the vast majority of basic research in this country is funded by the federal government, these recommendations focus on government funding organizations. Nevertheless, many of the key points apply to private sources of capital as well.

Focus on Important Problems

Important and also interesting problems are all around us. There are many examples in health care and medicine, the environment, transportation, telecommunications, and defense. Chemistry has and will continue to have a major impact in these fields. Today's problems are inherently multidisciplinary and a challenge to our current university structure. This challenge must be met by any funding organization.

For example, in the area of defense the world is changing rapidly. No longer are we fighting the Cold War, and new threats are emerging everywhere, as evidenced by the increase in terrorism, the proliferation of chemical and biological weapons, and the advent of information warfare. At the same time, our adversaries have access to our latest technologies and there is increased pressure on the military to cut costs. It is very clear that the DOD is no longer the technological leader in such fields as advanced electronics and information technology and that it will never be the leader in the burgeoning field of biotechnology. These ideas do not come from any classified government documents but from reading newspapers like the *New York Times* or the *Washington Post* and from watching CNN.

Against this backdrop, chemistry can play a key role. Chemistry and chemical engineers can develop new materials and actuators for unmanned and robotic platforms and can use biomimetic or bio-inspired principles for sensors to detect the chemical signatures of land mines, chemical weapons, or biological weapons. They can also control micro- and nanostructures to improve ballistic protection, develop new therapeutics to counter the effects of emerging chemical and biological warfare agents, and develop self-healing materials to repair our aging platforms. These are but a few of many possible important problems in the defense arena where chemists can not only perform great, intellectually challenging science but also make a real impact on people's lives. Similar lists can be made for many other fields.

By understanding the broad market needs irrespective of any individual technology, one naturally focuses on important, interesting, and technically demanding problems. The consequences of this include the following:

- In a university setting, research becomes more relevant to students and staff. Problems are inherently multidisciplinary, and students learn naturally from their colleagues in other departments.
- There is a demonstrative value to society that enhances the ability to move basic research rapidly into the marketplace and therefore enhances the potential for real economic value creation.
- Increased government and industrial investment in research and research infrastructure allows access to more advanced research tools.

Because the field now has more impact, there is a natural, positive feedback loop causing the field to grow and become more important. Witness, for example, the growth in funding for the National Institutes of Health.

Keep the End in Mind

The specifics of the end-game approach to research management were discussed above, and the advantages are many:

- Investment is more focused, and return on investment is more easily viewed and measured.
- Value to society is more clearly demonstrated because the goal is well defined from the start. Thus government funding can be expected to grow.
- Correlation between investment and scientific and economic progress becomes clearer. This is especially true when proactive program management is tied in (see below).

- The chemical industry can become more engaged in basic (university) research because outcomes are more relevant to industry, leading to more jobs for students and new and faster innovation.
- Research and teaching integrate naturally because not only do students see the relevance of their work, they also are being trained for the "real world."

Empower Funding Organizations

In order for an "end-game" research management program to be successful, funding organizations that play more proactive roles are required. For example, funding organizations must provide a clear need, priorities, and well-defined goals to both their constituents and their customers. The staff of the funding organization must understand both government and societal needs and be able to mix strategic (global) and tactical (directed) research and development. This means a mixed-risk approach should be used, combining appropriate amounts of basic research, applied research, development, and demonstration. The funding agency must support teaming and the effective use of scarce experimental resources through, for example, partnerships with national labs, not-for-profit organizations, and commercial companies. The government should encourage the use of corporate or individual donations of funds and equipment to tie the public and private sectors closer together and to enhance the training of students in state-of-the-art facilities. Funding organizations should also work closer together to gain critical mass and to minimize any duplication of effort. In order to make appropriate funding decisions in a timely manner, government funding agencies must optimize the use of multiorganizational panel review, peer review, and the intelligence and "gut feel" of individual program managers, whose technical judgment and expertise should be valued.

Since technology is not standing still, government program directors must have both the technical and the fiscal flexibility to review and change funding between and among scientific and engineering disciplines. While these decisions are difficult to make, leaving the tough choices to technically unqualified bureaucrats and legislators will not be in the best interest of our future. Similarly, resource allocation such as spending on equipment versus salaries, or funding of "big" science versus "small" science, also should be made by those most technically qualified. By making connections between research groups and fostering an atmosphere of collaboration, government program managers could also provide a very valuable service—enhancing technology transfer. This does not entail commercialization of technologies per se, but it ensures that there is a free flow of knowledge from those who generate it to those who will ultimately need it. Finally, if research is not progressing appropriately and fields are "getting stale," program managers should have the freedom and ability to terminate projects and invest the resulting funds more productively elsewhere.

CONCLUSIONS

While these three recommendations may appear to be somewhat radical, organizations such as DARPA have used these techniques to very effectively move technology rapidly into the hands of those who need it most. Although these approaches may not work for everyone, they have been proven over time to be very effective. They require funding organizations to be proactive and to rely on the skills and judgment of their research managers.

DISCUSSION

Michael Schrage, Massachusetts Institute of Technology: I have actually followed a lot of DARPA's work over the years, and there is one question by which I am really struck. DARPA has always been a multidisciplinary agency, so how do you, as a program manager, strike a balance between the way a project is defined versus how you facilitate and translate between different disciplines?

Lawrence H. Dubois: It helps when you have money because it enables you to do a lot of different things. One example is at a principal investigator's meeting we will bring people together and have a series of lectures/discussions. People ask questions. Maybe initially you have the physicists asking questions of physicists and the chemists asking questions of chemists, but then we will typically do something a little out of the ordinary. For example, we will rent one of those boats that go out on the Potomac River from 5 o'clock until 10 o'clock at night. Everybody is on the boat at sea and you have to talk to somebody, right? There is food. There is drink. We have played paint ball games and laser tag. We do anything we can to break down barriers between people. That is just an example of what we will do to help people communicate, and it does take time.

Another example uses technology demonstrations, where every team brings their technology and demonstrates it. Each team will have a little booth or table, and the requirement is that somebody from your team has to go work with somebody else on another team to help set up their demo. It doesn't matter how simple the task; the goal is to help people talk to one another.

Finally, from a programmatic standpoint, one of the critically important procedures we used in the Defense Sciences Office of DARPA was that when we hired a new program manager we did not let him or her run programs in their area of expertise. For example, if someone is an expert in semiconductor processing, they were not running programs in semiconductor processing. They wouldn't do something totally foreign like medical applications on the battlefield, but it would be an area that is a little outside of their comfort level. It allowed a program manager to ask "stupid questions." In an unfamiliar field, one could ask "Why do you do it that way?" or "How do you do that, and what does that really mean?" If this were the manager's area of expertise, he or she would be "banned" from asking those questions. When you pull people outside of their comfort level, and the funds are there to back you, everybody wants to educate you. That is another way of breaking down a lot of these communications barriers.

Michael Schrage: One quick follow-up on that. You came up with the program's end points. What is the tradeoff between how rigorously you define the end points versus emergent specifications and emergent prototypes from the team you put together? How is that negotiated?

Lawrence H. Dubois: It goes back and forth. You have this goal sitting out there, but you can't reach the goal without a team. For example, the team says, "You know, we need more models. We've got these concepts, but the model is wrong. Larry, I have a suggestion. Could you find somebody who can help on the modeling side because we don't think we can reach the goals without it?" There is definitely iteration between the team and the program manager. However, in many cases I have seen program managers be pretty adamant by saying, "Okay, these are my goals. You go figure it out." It really pushes the team; if the team members don't figure it out, they may not get additional funding. DARPA truly pushes both its program managers and its contractor base.

Richard J. Colton, Naval Research Laboratory: Is your description of this program something that all of the offices of DARPA subscribe to? Was this methodology developed by DARPA when it was established, or is it the way to operate DARPA according to Larry Dubois? Also, previously DARPA was very much a bottom-up organization, but it is rumored that DARPA is now more top-down. Can you comment on that under the new management?

Lawrence H. Dubois: Sure. First, the Defense Sciences Office was very different from most other offices, partly due to the kind of people we hired. We brought in a much more technically diverse group of people than any other office. In addition, each office tends to take on the character of the office director. My philosophy was very much bottom-up. I would sign almost anything that anybody put in front of me as a funding document. I might argue with them. I might get them to rewrite it or change the scope of work, but ultimately I would sign it because you have to trust your program managers. That is very different than some of the other office directors who were very much top-down managers. In this case program managers tended to be an extension of his or her ideas. DARPA does change depending on the character of not only the program managers and the office directors but also the director of the agency. I think that the new director has made DARPA more of a top-down organization, and program managers execute what it is that he likes. Like anything else, however, the way DARPA is run will continuously evolve because the director will be there for only a few years.

5

Comments on the Advanced Technology Program

Mary L. Good[1]
University of Arkansas, Little Rock

The political ideology associated with the federal government's support of commercially oriented technology has been an enormous problem for the Advanced Technology Program (ATP) since the program's inception. This is despite the fact that the federal government has historically supported technology, research, and development in a variety of industries. For example, the federal government paid for the first demonstration of the telegraph. Research and technology development for aviation has been almost entirely funded by the federal government, starting during the First World War and continuing even throughout the Depression. This investment became an enormous asset during the Second World War. The aerospace industry continues to believe that it is the government's duty to supply that research base, and indeed, for the most part the government still supports research and development (R&D) in that industry.

The Internet was created through federal funding by the Advanced Research Projects Agency and was originally designed so that scientists working for the Department of Defense could efficiently communicate with each other. The project was later transferred to the National Science Foundation (NSF), which provided the backbone for the Internet. In a similar manner, the federal government has funded most of the research in the agricultural technology industry. With the history of government intervention in these industries, it is intriguing that ATP has received such unusual feedback with respect to its role in technology development.

ATP was initiated in 1988 by President George H. W. Bush amid a climate of domestic collaboration to fight international competition. The Japanese had become a threat to the U.S. economy because of their success in developing new manufacturing technologies. The government responded by passing a number of technology and innovation acts. These included the Bayh-Dole Act of 1980, which gave the rights to the intellectual property of government-funded research to universities, small businesses, and

---

[1]Mary L. Good is the Donaghey University Professor at the University of Arkansas, Little Rock, and serves as the managing member for Venture Capital Investors, LLC. Previously, she served 4 years as the under secretary for technology for the Technology Administration in the Department of Commerce, a presidentially appointed, Senate-confirmed position, which had oversight of the Advanced Technology Program.

eventually large businesses. The Small Business Innovation Research Act followed in 1982 to strengthen the role of small innovative firms in federally funded R&D as a base for technological innovation to meet agency needs and contribute to the growth and strengthening of the nation's economy. Semiconductor Manufacturing Technology, SEMATECH, a research consortium of the U.S. semiconductor industry began in 1987 in response to the economic downturn and international competition and is credited with returning the U.S. industry's dominance.

Legislation for the creation of ATP was included in the Omnibus Trade and Competitiveness Act of 1988. The idea was to provide partial funding for projects that were based on worthwhile ideas with commercial potential but that were at such an early stage of development that they were not likely to be supported by industrial firms. The projects were to be supported through the precommercial research stage but not into commercial development. Many of these projects paired small companies with large companies or universities.

ATP's first budget was approximately $10 million, but by 1992 its allocation had risen to about $60 million per year. That same year the Clinton-Gore campaign made ATP part of the Democratic candidates' science and technology plan, and in 1994 ATP's budget was over $200 million. Unfortunately, the support of ATP by Clinton-Gore gave the impression that it was a political program. When the House of Representatives and Congress changed party leadership, the program was vigorously attacked and the era of political debate over ATP's future began. Amazingly, the program still runs with a budget of approximately $200 million, although it is by far the most evaluated government program ever—it has been analyzed by economists, scientists, industrial participants, and politicians. Although some changes have been made to meet the demands of various ideologies, the program still exists.

About $61 million in new money will be available for research programs through ATP for fiscal year 2003. The remaining $140 million will be used for continuing grants from previous campaigns. The program has never been fully funded by the federal government; half of the funding is required to come from industry. While most of the participants are small and large businesses, universities have been major participants in ATP from the program's conception. There have been a total of 581 projects involving 150 individual universities. ATP has been a fairly sizable supplier of funding to universities for their research activities. National laboratories have also participated in ATP.

There have been many successes in ATP. Even the opponents of the program agree that it has been successful. The question then is, "Is the program needed?"

ATP funds early, precommercial, enabling technology research that cannot be funded by government basic science programs like NSF based on program rules and definitions. Private companies generally do not fund research on enabling technology because, as use of the technology spreads, it is difficult to recover the costs. Although enabling technology could be important for the country as a whole, an individual company will not invest in that technology if it is not driving the market at that moment. There is actually very little private-sector capital for early-stage technology research and development. Venture capitalists want a prototype and a product market with a known size. Venture capitalist and founding partner of Morganthaler Ventures David Morganthaler has stated: "It does seem that early-stage help by the government in developing platform technologies and financing scientific discoveries is directed exactly at the areas where institutional venture capitalists cannot and will not go. In the analogy of the horse race, the role of the government can be to improve the blood lines of the horses and give them some preliminary schooling, but it is not buying the horses."[2]

---

[2]David Morganthaler. April 2000. Assessing technical risk. Unpublished presentation, National Institute of Standards and Technology.

The private-sector venture money also has a tendency to "follow the herd." As a result, certain technologies are funded preferentially. For example, in the 1980s, anything related to biotechnology was funded. In the past decade it has been information technology. Materials research has been ignored mainly because it involves extended development. The demise of the Department of Defense's funding of technology since the Cold War has also contributed to the lack of available funding sources. Lastly, the corporate culture at large companies routinely causes termination of long-term precommercial research and favors projects that will make the company $100 million in the very near future. For this reason, virtually all of the in-house venture funding programs have folded. Clearly, a program such as ATP, which funds precommercial research that otherwise would not be pursued by the private sector, is needed.

The opposite viewpoint is also heard. Some believe that the government should not fund technology research and that it should be left to the private sector. Government funding of research also gives large companies access to government subsidies. Although some contend that the government should not pick "winners" or "losers," the NSF has always selected winners and losers for its research grants. In an attempt to create an evaluation system that was nonpolitical, the sponsor of ATP, the National Institute of Standards and Technology (NIST), has used reviewers without conflict of interest but with a real understanding of the technology. In fact, most of the reviewers have been recently retired scientists and engineers from universities or government agencies (except the Department of Commerce). However, the selection process becomes difficult when a small company involves its congressman and adds a political aspect to the selection process.

The eventual commercialization rate of the projects in ATP is high. This is essential for the program's survival since funding for ATP is debated on a yearly basis. Unfortunately, this is most likely due to selection criteria that favor lower-risk projects at the expense of high-risk and high-payoff projects.

The federal government's research and development portfolio contains fundamental research that is not targeted toward any foreseeable commercial use, applied research, or development. In general, applied research is designed to provide answers to specific scientific and technical questions needed to carry out certain governmental missions in defense, energy, space, the environment, and underlying national interests in the commercial sector such as standards and metrology. However, there are areas that are neglected by the federal government and corporate research programs. The ATP is an attempt to provide incentives for the development of new technologies before the usual market forces will focus on them. Not only will this provide significant additions to the nation's technology pool, a program like ATP has the potential to create new economic growth areas and provide opportunity for entrepreneurs in any geographic location.

The ATP fills in the research gaps between the mission agencies and fundamental research agencies. It has a well-defined, worthy purpose and is by far one of the best-run programs. I would like to see ATP continued and stabilized at approximately one-half billion dollars a year, which will bring out the best pool of applicants from industry.

DISCUSSION

Robert A. Beyerlein, NIST: I represent ATP and I want to say that I very much appreciate Dr. Good's excellent even-handed perspective on the program and rather insightful recommendations for change.

I only offer one comment about the current competition and that is what NIST and ATP are advertising. If you get in your proposal for this current competition by June 10, and if you win, you are recommended for funding in 2002. However, if you get it in later, any time up to September 30, the chance for funding most likely will be received from 2003 funding.

Mary L. Good: NIST and ATP expect to receive enough proposals to fund $60 million by June 10.

Robert A. Beyerlein: Right. As you stated earlier, some of that $61 million will be used to fund some projects that will be announced in the near future. These projects came in the later stages of 2001.

Joseph A. Akkara, NSF: I was very interested by your portfolio of management prescription. Who manages the portfolio?

Mary L. Good: That has always been the problem. Who manages the portfolio? Many of us have spent a fair amount of time thinking about the government's funding of R&D. It would be wonderful if somebody moved up to the plate and said, "We really need to look at our portfolio and to know what we are doing and have some minimal oversight of it." I wouldn't recommend detailed micromanagement; that is not what I am talking about. Whenever a portfolio is assigned to the agency, it is the agency's to manage. However, I had always hoped that the Office of Science and Technology Policy (OSTP) would step up to that plate and manage the whole portfolio. OSTP is one of the few groups that can.

We currently have the concept of an interagency group that advises the science adviser; this idea has survived from the first Bush administration. If the right representatives from the agencies in that interagency group are involved in addition to OSTP, they would be in a position to look at the portfolio, evaluate it, and get feedback from a group like the President's Council of Advisers for Science and Technology. The council is able to give some outside review of what path internal documents suggest.

I understand the political difficulties of playing around in certain committees' territories here, but sooner or later we need a serious evaluation because we are not necessarily using our R&D dollars for the biggest bang for the buck at the moment.

We had a conversation at lunch about the Department of Energy (DOE) stating that it is now the source of the major facilities that people use. That is where it ought to be. You can't reproduce facilities like the light sources on university campuses, nor can you reproduce them in industry, but such facilities ought to be part of that portfolio. DOE is providing the infrastructure for the country and that ought to be part of the country's management portfolio.

Joseph S. Francisco, Purdue University: What is PCAST?

Mary L. Good: PCAST is the President's Council of Advisers for Science and Technology, and it consists of approximately 24 members from industry, academia, and nonprofit organizations. PCAST has some very significant people on it, and they could be used as a sounding board for whatever kind of organizational structure and oversight issues the government might handle.

Larry Dubois, SRI: ATP is one giant step toward the commercial side, away from the type of funding that NSF would give and away from the intelligence community's programs and the Army's venture capital. Would you comment on this trend of the government getting into the venture capital world?

Mary L. Good: That is an excellent question. I would remind you that in many ways if I look at how deeply the government was involved in the aerospace industry, ATP is nowhere near that level of penetration. We are a long, long way from driving the industry with any of these programs, but my view is that the portfolio for the government ought to have this balance. If you take ATP, add about one-half billion dollars, then factor in DARPA at about that same number—do you know what the DARPA budget is?

Larry Dubois: About $2.3 billion.

Mary L. Good: The DARPA budget itself?

Larry Dubois: The DARPA budget is about $2.3 billion also.

Mary L. Good: This is all external?

Larry Dubois: Yes, but if you add all of it up, the total amount of money going to the early stage of technology research is a fraction of what we spend in R&D for the government. As long as it stays at 5 percent, it would be an excellent investment. I wouldn't even object to going as high as 10 percent, and it is nowhere near that at this stage of the game.

The stuff that the Department of Defense (DOD) is now asking for is commercial because it goes beyond what ATP does. DOD has done that before. DOD no longer has commercial contracts that have money set aside for research. In the past, if you got a contract, you also got a research budget to go with it. DOD doesn't do that anymore. If you look at the total amount of DOD grant dollars given for research, it is less than it used to be, not more.

Hank Whalen, American Chemical Society: I have two questions. First, of the $200 million roughly that is appropriated for this year, what percentage would the chemical industry have of that?

Mary L. Good: I don't know. It would be small though. The chemical industry, for various and sundry reasons, has not participated in ATP to the extent that it might. On the other hand, when I say that, I am being somewhat disingenuous because the materials scientists have participated. Do I count them as chemists or not?

Hank Whalen: It depends on to whom you are talking.

Mary L. Good: Exactly. It depends on to whom you are talking. There have actually been three or four chemical companies involved. DuPont and a couple of small companies did some very innovative work on utilizing some plastics for commercial opportunities that had never been looked at before. Nevertheless, participation in ATP by the chemical industry has been rather small. The chemical industry is very proprietary, and many chemical companies are unwilling to go through the review process that NIST requires.

Hank Whalen: Newt Gingrich and Bob Walker were very negative toward ATP. There doesn't seem to be anybody that is that negative right now.

Mary L. Good: I think that is probably correct. As you know, ATP was funded again last year, although it was not in the president's budget. If I were a betting person, I would bet it will get funded again at about the same level whether it is in the president's budget or not.

There are enough people who believe that the program actually does something significant and want to support it. On the other hand, they are not willing to get out and push for big increases, particularly at a time when the budget is pretty tight.

Mary L. Mandich, Bell Labs: Can you give the rationale behind backing smaller companies as opposed to large companies? Then would you care to defend that?

Mary L. Good: Do you mean that small companies are better than large companies?

Mary L. Mandich: Yes, in terms of getting funds from ATP.

Mary L. Good: Actually I would not defend it very vigorously. Let me put it this way. The reason I wouldn't is that some of the very, very best projects that ATP has funded were large companies working with small companies and with university experts. That combination has made some of the very best projects in terms of real impact and commercialization. There is a reason for that. The large companies, in many cases, have the expertise that can actually help that little company make that technology move, while a small company just doesn't have the expertise to do this at the early stage of technology research.

We are not talking about just building a prototype and commercialization here because ATP stops at the first prototype.

Mary L. Mandich: What is the rationale, and would portfolio management of some sort help remedy this issue?

Mary L. Good: The rationale truly is almost ideological. People don't believe that large companies should be subsidized for research they believe the companies should do anyway. There is also a myth that small start-up companies should get some help. But in terms of the value of results, for example, DARPA does not, should not, and probably never will eliminate large companies. Even the new ATP rules will still let large companies participate, but there was a thought process that said that the American economy was built on start-ups. What people forget is to look at the whole infrastructure.

Most of the small companies that have started up in this country and that have done well were either started by people who came out of big companies, learned how to do business, and ended up with the technology the company didn't want, or the company decided the technology was too small to worry about and allowed it to be taken outside. People see all these small companies, but they don't understand that those wouldn't exist without training and background from larger companies. I just went to a big venture capital forum recently and was amused because of the five or six companies that were chosen to highlight for the venture capitalists, every single one of them had a CEO with gray hair.

Peter Koen, Stevens Institute: Several years ago I looked at the amount of money Japan was putting into its Ministry of International Trade and Industry (MITI, in 2001 reorganized to the Ministry of Economy, Trade and Industry) and compared it to ATP. I expected ATP to be funded to the tune of at least a few billion dollars, but it is not. What is wrong? How are we sending out the message so that our congressmen and our representatives don't understand the importance of funding the basic research?

Mary L. Good: First of all, I don't think those of us who have looked at MITI would argue that we should do what MITI does. If you look at MITI's success rate, it is not what I would advocate. However, ATP ought to be focused in those areas that the market will not fund and does not fund at the moment. Technology research is a legitimate business and that is one of the problems. When people talk on Capitol Hill, they talk about science, and we are not talking about science per se. There is science involved, but we are talking about technology research. It is partially what engineers do, among other things, but technology research still requires a fair amount of research. I think we should have some mechanism that allows people who have good ideas in this arena to have a way to get off the ground. This is a very difficult area for even angel investors to be involved in because the timescale is too long.

It takes too long and takes too much money. You can get a few of them to do it, but it is not a venture capitalist's normal undertaking. Today, biotech and information technologies are fundable but not materials science. I can guarantee it. Somebody needs to fund materials science because in many ways materials are the basis of all the technologies that follow.

I liked Elsa's remarks this morning because it is clear that without the materials research there is no progress. That is true across the board in many other areas, but only a few of the big companies will still fund materials research. They expect their suppliers to do it, but their suppliers don't get enough margin to fund it anymore.

Participant: Will the myth continue that this is what companies should be doing rather than our government?

Mary L. Good: It is an interesting thing. ATP was born out of a crisis. I would predict that within the next 3 or 4 years we will have another crisis of some kind. Then ATP will get a new lease on life and perhaps a different title, but that is the way we work in this country. We respond to crises, and in a sense that is the answer to the question about what the Defense Department is asking in terms of commercialization for security technology. Something like national security is a crisis, and agencies respond as they ought to. Hopefully, the country's policymakers will find a middle road between investment for the future and crisis management.

6

What Have We Learned from Hot Topics?

James R. Heath[1]
University of California, Los Angeles

This chapter describes two high-profile projects—one that involves an institute that I was involved in building, and the other is a scientific program that I have led. Through these projects, I will address issues such as funding for long-term research, high-risk projects, the bonuses and pitfalls of industrial collaboration, and technology transfer.

It is always logical to put science on top, and so I will talk about that first. The scientific program began several years ago when I visited the Hewlett-Packard (HP) Laboratories to help my friend Stan Williams set up a wet-chemistry effort there and to work on a collaboration involving the fairly esoteric task of performing transport spectroscopy measurements of two-dimensional solids of quantum dots. This project was funded by the National Science Foundation's (NSF) Grant Opportunities for Academic Liaison with Industry (GOALI) program, which also included Paul Alivisatos. I will come back to the nature of this funding later. In any case, Stan and I soon began to talk about other things, such as whether it would be possible to utilize self-assembly, or something like it, to build a computing machine. This was a problem that I had thought about a few years before when I was on the staff at IBM, but it was one of those thoughts that just didn't go anywhere because I didn't have sufficient "fuel" to run with it. Stan and I had, in the back of our heads, the concept that such a machine might be extremely energy efficient. Consider, for example the cost of switching 1's into 0's and back again. Rolf Landauer and Richard Feynman thought about this concept many years before, so we had very smart people to learn from.

A 1 and a 0 have to be energetically different from each other or you can't tell the difference between them; therefore, it costs energy to do this switching process. If a calculation is done by switching the 1's and 0's, there are going to be entropy costs as well. In any case, a 1-electron (quantum state) switch appeared to be an attractive option for minimizing the energy of this process. To be robust, the energy difference between a 1 and a 0 should be about 20 to 50 $k_B T$, and we wanted T to be room temperature. By simply considering a particle-in-a-box model for calculating energy level spacings, it

---

[1] James Heath is currently professor of chemistry at the University of California, Los Angeles, and director of the California NanoSystems Institute, formed by California Governor Grey Davis in December 2000. He was previously a research staff member at the IBM T. J. Watson Research Labs.

becomes apparent that this switch is the size of a molecule. Nevertheless, Stan and I didn't have much of a clue as to how to make a molecular switch, much less a large-scale circuit of such switches, as would be needed for a computer. Thus, we started with nothing but an aspiration to build this computer. To be honest, I wanted to build the computer simply because it represented a significant challenge, and we clearly did not know how to do it. I thought that if we did build it, it didn't even have to be useful to be worthwhile. We would very likely learn some wonderful science along the way, and we might even discover things that were useful. That has, in fact, happened, and I will return to this point.

First, let me talk about nanotechnology and how this project was funded. The project was started before the National Nanotechnology Initiative, and funding for ill-defined (call it exploratory), cross-disciplinary research was (and still is) hard to get. Thus we submitted a proposal to NSF's GOALI program that actually was fairly focused. The great thing about NSF is that the agency doesn't really care what you do with the funding it gives you, as long as you perform high-quality science. NSF also supports long-term projects, but this is a concept that must be continually retaught to our taxpayers (and even to NSF program monitors on occasion). The advent of the National Nanotechnology Initiative and similar programs has begun to educate the public about the importance of long-term research. Nevertheless, nothing helps garner public and federal research support more than pointing to legitimate commercial products that have come out of government-funded research. This is understandable and as it should be, since there must be some level of payoff to the people who pay for it. However, consider the case of nanotechnology, in comparison with biotechnology. Research in biotechnology has stabilized only in the past few years to the extent that there are a number of companies making money. Figure 6.1, which was put together by my colleagues Michael Darby and Lynn Zucker at the University of

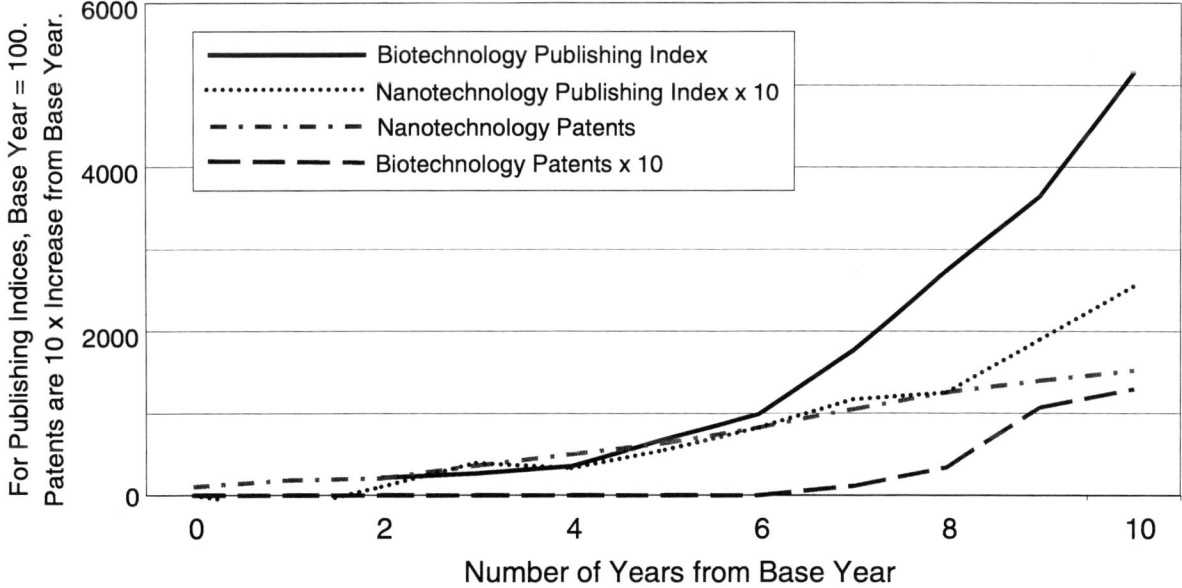

FIGURE 6.1 The time lag from an emerging science's base year to the time it becomes productive causes a delay in the financial returns on investment. The base year for biotechnology is 1973 and the base year for nanotechnology is 1989. Courtesy of Michael Darby and Lynn Zucker.

California, Los Angeles (UCLA), shows a similar rise in the number of publications and patents for nanotechnology and biotechnology, looking at the number of years from the field's base year. Those base years are 1973 and 1989 for biotechnology and nanotechnology, respectively, and thus nanotechnology should become a major commercial sector in 10 to 15 years. This can only happen if we continue to invest and nurture this field. Private money (venture capital) will not be available, because it is too early. Therefore, nanotechnology will not yield returns without sustained federal funding, but it will likely earn wonderful returns if we have the willpower to stick with it. This is an argument that must be made time and time again to our government leaders and to the people who vote for them. This is true for nanotechnology and it is true for any other basic science initiative our nation undertakes.

The NSF GOALI grant that we received allowed a small group of people with different backgrounds to take the time to come to a common understanding on a particular issue. This was really critical, or the project would not have gone anywhere. Let me say a few words about how difficult it is for people from very different backgrounds to come together. In addition to our original quantum dot project, Stan and I began to interact with Phil Kuekes, a computer architect at HP, because Phil had done some interesting work in terms of building a molecular computer. Stan and I wanted to learn from Phil what he knew, and he was anxious to learn some physics and chemistry. We sat together many times for 4-hour sessions trying to learn each other's language. I remember a few times walking out of one of those 4-hour sessions with only a common noun.

Phil had built a wonderful machine called Teramac, which consisted of circuits that were laid out in a "periodic" (as a chemist I thought "crystalline") arrangement, and it worked perfectly in spite of having a quarter million "defective components" (again I translated this into finite chemical reaction yield). A Pentium computer chip, by contrast, consists of very complex circuit arrangements with no defective components. Stan, Phil, and I used the architecture of Teramac to develop a model for our computing machine. We published this architecture and its implications for nanotechnology in *Science* in 1998.[2] We received a lot of press from that paper, and we also got the attention of the Defense Advanced Research Projects Agency (DARPA). Unlike the NSF, which gave us the freedom to explore but not a lot of money to explore with, DARPA made us focus on a set of deliverables but gave us a lot of money to produce them. We promised DARPA a 16-bit molecular electronic-based random access (reconfigurable) memory circuit within 2 years. To be honest, this deliverable scared us, but we made very rapid progress. In 1999, just prior to the start of the DARPA program, my collaborators and I physically demonstrated that molecules could be useful for simple logic circuitry, and in 2000 we reported the first truly reconfigurable switches based on these molecules using a one-electron process. By the next year we had a 64-bit random access memory working, and this work was cited in December of 2001 as *Science* journal's Breakthrough of the Year. In 2000, defect-tolerant architectures were placed onto the Semiconductor Industry Association Roadmap—something that was catalyzed by our Teramac paper. In 2001, molecular electronics was also placed on the Roadmap, although as a far-term research goal. Recently, we have demonstrated patterning techniques that allow us to make memory circuits that are similar to the 64-bit memory we delivered to DARPA but at a density approaching $10^{12}/cm^2$, which is, in fact, actually beyond where Stan and I had originally intended this project to go.

This research has received a lot of media attention. I had previous experience with the press through my work on the discovery of C_{60}, as well as some other research we had published in the mid-1990s out of my UCLA group. Nevertheless, I did not realize at the time that the press would take one short quote

---

[2]J.R. Heath, P.J. Kuekes, G. Snider, and R.S. Williams. 1998. A Defect-tolerant Computer Architecture: Opportunities for Nanotechnology. *Science* 280:1716.

from over an hour of conversation and use it as the whole basis of an article. Some of the quotes put quite a bit of pressure on us. For example, one of the quotes included the phrase "a thousand Pentiums on a grain of sand," which made it into the State of the Union Address a few years later. When this quote was made in 1998, the highest possible bit density was about 10^8 bits/cm^2. If a Pentium chip is about 1 cm^2, then 1,000 Pentiums yield about 10^{11} bits. A grain of sand has a volume of about 8 mm^3. Since computer architectures are typically one-dimensional, the 10^{11} bits can be stacked using four layers. The required bit density to have 1,000 Pentiums in a grain of sand is about 6.25 (10^{11} bits/cm^2 per 2-mm-high stack, which is nearly equal to the volume storage of the human brain. This is actually the density of devices that we are working on developing right now, although we only have a few tens of thousands of devices in our largest circuits at this high density. Furthermore, making integrated circuits (logic and memory talking with each other and with the outside world) represents a significant step in complexity beyond where we are working now.

Why does HP support this project? Stan Williams leads for this project at HP, which is nearly the only blue-skies project that HP does fund. The assistance that they get from DARPA is critical to the program. The culture at HP is like any other big company in the information technology business in that any one of HP's products and services will become obsolete within 3 years. The vast majority of its resources are therefore spent on reinventing themselves every 3 years, and HP pays very little attention to technology that clearly cannot be developed within that time frame. Although Stan has kept the HP ship fairly steady, working with HP has, at times, been a struggle. Largely because of Stan's efforts, we have been able to hold HP's attention thus far. This has given us benefit with DARPA as well. The very fact that HP has a strong interest in molecular electronics has given a significant stamp of approval to DARPA's goals for this program. For example, DARPA Director Tether has visited HP labs and been briefed on our molecular electronics program. He has not visited UCLA, Harvard, or other academic labs that are significant players in this field.

We have collectively received several million dollars in grants, which have been split roughly 50/50 between HP and UCLA. We have six jointly owned patents that have not only been filed but have been issued over the past few years. An unusual aspect of our collaborative agreement is that both UCLA and HP retain full ownership privileges for intellectual property that is jointly developed, regardless of the relative effort each party has put forth. It is unusual that UCLA retains complete ownership privileges (as does HP) through our joint arrangement. We also filed about six other separate patents during that same time period. We have published six joint papers and about 20 other papers; several of these papers have received international press coverage. This type of publicity is clearly of value to HP. Its efforts to publicize some of this work have been remarkable when compared with what universities do, but are probably not so unusual if one tracks the positive impact that such press releases have on HP's stock price. IBM operates similarly.

This program has accomplished much, but it has a long way to go before it produces useful technologies. Probably the most promising avenue is to create energy-efficient (our first goal) computational platforms that are substrate independent. This is a unique niche for molecular electronics that is not easily met by any other information technology platform. However, out of this work have also come avenues for fabricating very high frequency (GHz) resonators (helps your cell phone talk to your computer), ultra-high-density molecular sensor arrays for interfacing with the molecular language and the timescales of living cells, and several other wonder discoveries. I believe that these discoveries alone will more than justify the federal investment that has gone into this program.

Now let me talk about my high-profile project. For the past $2^1/_2$ years, I have been involved in the California NanoSystems Institute (CNSI), which is a $100 million program funded by Governor Gray Davis, and which is matched by $200 million from nonstate funding sources. Our objective is to be the catalyst to make Southern California the birthplace of the nanotechnology industry. Our mission, as stated by Governor Davis, is to keep California's technology industry thriving for the next 30 years by following these guidelines:

- Science—lay the foundation
- Technology—take it the next step
- People—train the work force
- Corporate contracts and start-up companies—facilitate cooperation
- Intellectual property (IP)—find new ways to transfer technology

In my experience at the University of California (UC), managing IP has been a very difficult process. When companies had shown interest in forming a co-development agreement with UC, more often than not they ultimately gave up due to the barriers involved in the process. Some faculty have left UC for start-up companies as a result of the university bureaucracy's tight and conservative policies regarding IP. As a result of the lack of success, Governor Davis gave CNSI a specific directive to manage IP differently, in a more liberal and open manner.

It is of maximum benefit to UC to have companies spring from its research labs but to have the faculty remain to invent and discover yet again. This requires a facile IP program that works for the faculty and is responsive to the needs of industry, venture capitalists, and others. Consider what is involved in a technology transfer process, especially in a high-technology area such as nanotechnology. A patent, while sufficient to get venture capital interests, is not by itself very useful. Knowledge transfer from the academic lab to the start-up or existing company is a critical component, and all of this—IP licensing, knowledge transfer, and so on—must be done rapidly to maintain a competitive edge. At many universities, including UC, this is nearly impossible to do.

Consider the following example. Several years ago my colleague at UC, Mike Phelps, devised a demonstration project that was to illustrate a new way of managing IP. His goal was to build something known as the LA Tech Center, which was a research arm of a company known as CTI Molecular Imaging. CTI is a publicly traded $1.5 billion company that makes PET scanners and is very successful. CTI came out of Mike's labs (Mike is the inventor of PET). Anyway, this project required two deals to be made. One was done by hiring a private lawyer who had to work with a single individual at the UC Office of the President in Oakland (no committees and high-quality legal assistance). The other was done through the regular channels and required a committee for the UCLA IP office, a committee for the president's office of the UC system, low-quality lawyers assigned by the university, and nearly complete exclusion of Mike in the negotiation process.

As of this writing, the LA Tech Center has been up and running and profitable for $1^1/_2$ years, because the first (demonstration) deal was done. The second deal is not yet done! The goal of the LA Tech Center is very interesting. It relates back to the issues of patents and knowledge going hand in hand to transition IP out of the university and into the commercial sector. Figure 6.2 shows the setup of a research facility in the LA Tech Center, a unique collaboration between academia and industry. At one side is basic research that reaches back to UCLA and CNSI, with a scientific motivation and fundamental scientific approach. The commercial motivation with the applied science and engineering approach is kept separate on the other side of the building. The boundary separating the scientific from the commercial is mobile according to the project's need. While the money that runs the LA Tech Center doesn't change sides, the people do. This is an extremely valuable aspect of tech transfer.

Based on ideas like this, and on my own frustrating experiences with the UC office of IP, I drafted a charter for the CNSI that does several things. First, the CNSI answers to a board that, while having three UC representatives, also has eight other members. Second, it gives the CNSI faculty their own

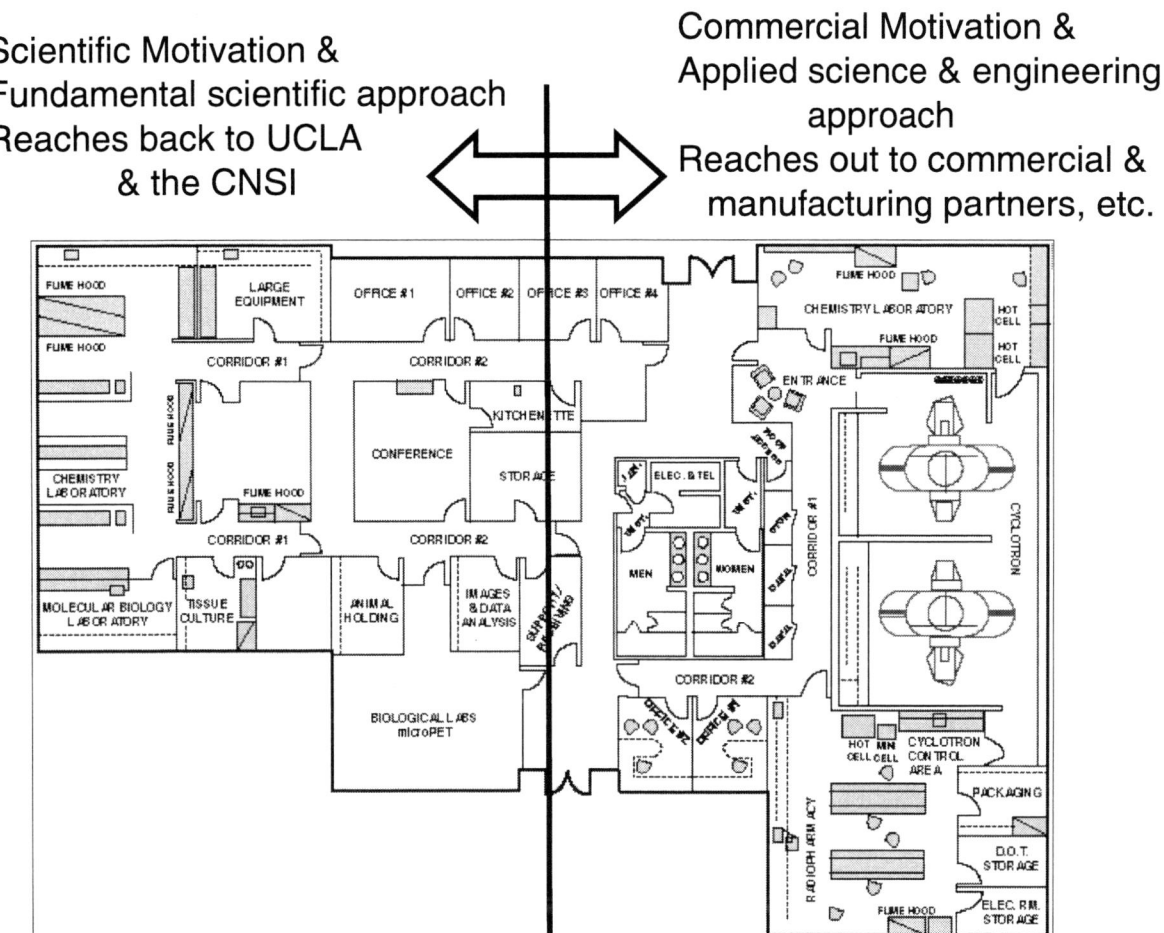

FIGURE 6.2 The commercial side of the LA Tech Center is highly leveraged by the scientific side (100:10); the boundary separating the scientific from the commercial is mobile according to the project's needs.

power over membership and building space. Finally, it gives CNSI the authority to pursue its own IP program. This means that CNSI can retain independent council and it doesn't have to move through the slow beaurocracy of UC. Furthermore, the director or co-director has signature authority on intellectual property. Speed is also a crucial factor in IP transfer, and the exclusion of a committee-based decision process has allowed CNSI to maintain a one-month time line for its operations. CNSI is a joint UCLA/ UC at Santa Barbara Institute, and both Chancellors Carnesale (UCLA) and Yang (UCSB) have signed this document, as have all of the CNSI faculty. The UC Office of the President has been reluctant to sign it, but I believe will eventually come around.

Nanotechnology needs long-term funding, which currently does not exist in the private sector. Corporate research labs occasionally fund such projects, but the prospects for corporate involvement in long-term research projects have diminished greatly over the past decade. Other private funding, such as what is represented by venture capital firms, is also not appropriate for long-term, high-payoff projects. For these reasons and to maintain the U.S. high-technology industrial sector into the foreseeable future, it is vital for research support to be available through programs like NSF's GOALI, DARPA programs, the National Nanotechnology Initiative, and institutes like CNSI.

DISCUSSION

Mary L. Good, University of Arkansas, Little Rock: I am very impressed with your presentation. Are there other areas you believe to be as important as nanotechnology?

James R. Heath: One area that I think is extremely important and of national interest is detecting and responding to an unknown pathogen. If something is released in the environment today and people begin to die, there is no way pharmaceutical companies will be able to develop a drug in sufficient time to help. If you look at the physical problems involved, diagnosis is already a very difficult problem, and rapid pharmaceutical development is virtually impossible. Although currently no one knows how to diagnose a pathogen and develop a pharmaceutical response within a few days, there is no reason why it cannot be accomplished. People are beginning to think about that very seriously. I am developing a research program on detection with Lee Hood of the Institute for Systems Biology and Mike Phelps from the UCLA molecular and medical pharmacology department, for which the idea is to go from molecules to patients. There are a few other people who are beginning to do this as well. As medicine becomes a molecular world, people will begin to understand the mechanisms of disease. They begin to understand how to better use that knowledge, but that type of technology will be very hard to push forward without some sort of government assistance. In fact, if the Department of Defense was to develop a program along these lines with an eye toward national security, it would probably involve some combination of nanotechnology and biotechnology.

Kenneth A. Pickar, California Institute of Technology: I really want to compliment you because everyone knows the University of California is user hostile with respect to commercializing technology. It is not just the professors who hate it; it is also people on the outside. The best revolutions in technology transfer are still coming from people like you who find innovative solutions.

James R. Heath: Fighting this battle took too much out of me.

Kenneth A. Pickar: It is a lot harder when you are fighting upstream. Mary Good mentioned the problem of too much hype about energy in the 1970s. We all can discuss what traveling to the moon in the 1940s would have been like if we didn't have the technology to do it. It would have been an enormous flop as well. However, what I see as a major danger is not the potential of nanotechnology but all this horrific hype that could cause another biotechnology crash as it did back in the 1980s.

James R. Heath: You are right. At present, all of our technologies are based on "low information content" systems, such as small molecules and atoms, or periodic solids. The nanotechnology world is interested in "high information content" macromolecules as the functional units. We are in the process of learning how to incorporate form, function, and activity at the macromolecular level into nanotechnology. We are not certain how to do that yet, but I think it is a world-changing approach to science and technology. It is a very slow process, but we must always stress the excitement of this field while warding off immediate gratification. I would like to refer back to the previous comment about the national technology road map of semiconductors. The very fact that there is something called a "road map" dissuades students to pursue this. They think you look at the map and you do the prescribed research. That is not very exciting, but there is a world here such as what I described and in other areas like this where there is actually quite minimal knowledge and discovery science is happening at a rapid pace. The route toward developing nanotechnologies from this science is very long term, and so the excitement must last. A cautionary note about too much hype is also equally important.

Robert J. Bianchini, Colgate-Palmolive Company: We just signed an agreement with the University of Michigan to get into antimicrobial nano-emulsions to control anthrax and other diseases. We had a lot of problems getting IP results. In the end, the university did set up a spin-off company. If there is any way that IP issues can be resolved with universities, it certainly would help increase the rate of the innovation process. Our company wanted to protect itself, and we needed the agreement worked out before we could move forward.

James R. Heath: When a university lawyer and the committee behind the lawyer are arguing with IBM, the opponents are unequal. UC runs from that fight, and things get lost in committees as a result. Now UCLA (through the CNSI) offers some flexibility to allow faculty to obtain good, qualified counsel. In my opinion, this flexibility actually pays tremendous dividends. If a faculty member and a company want to make a deal with the university, policies are generally not so restrictive that they keep that from happening. It is the practice that is so frightening.

Joseph S. Francisco, Purdue University: How did you manage to get the University of California to allow you to agree with its policies while in turn taking a very different route in terms of practice?

James R. Heath: We have a bit of a luxury in that the policies at the University of California are actually very flexible. They have been built on top of each other over many years and just because there is a new rule doesn't mean it negates an old one. So one can do anything. However, the university's attitude is "I don't want to get sued, and our invention is worth $5 million." The attitude of the company on the other side is that your invention is worth 10,000 shares of start-up stock, the cost of the lawyer, and the cost of your IP filings. The truth is probably in between, but it is closer to 10,000 stock shares than $5 million. In terms of the CNSI charter, it wasn't very difficult to convince our chancellor that the process for translating IP to the commercial sector was broken. The CNSI also had significant weight in that a large fraction of the UCLA/UCSB inventors were part of the CNSI. However, the CNSI has only about 30 faculty total, so the absolute number is very small. Consider this argument that we gave our chancellor: UCLA is third in the country in terms of bringing the university revenue, and our IP office still runs in the red, and it has never come close to running in the black. In fact, if you were to factor in the cost of replacing faculty who leave UC to form companies, our IP office is in really bad shape. So that already is a pretty fine argument that there is a problem, and it is one that the UCSB and UCLA chancellors agreed with.

7

Industrial Innovation with External R&D Programs

Francis A. Via[1]
Fairfield Resources International

It is a privilege to be part of this Chemical Sciences Roundtable program with the National Academy of Sciences. The goal of this presentation is to demonstrate that external research programs with collaborators at universities and national laboratories can accelerate the industrial research and development (R&D) process. The essential criteria to realize this goal will be considered and illustrated with several examples that led to accelerated commercialization and some that did not.

Participating in this workshop has given me the opportunity to review Akzo Nobel's U.S. experiment with innovation over the past 12 years and to reassess the metrics for external research programs. I would like to provide an overview of the strategic value of external research programs using specific examples from Akzo Nobel supplemented with additional case studies from Dow and DuPont. This presentation will attempt to demonstrate that these collaborative research programs have become a critical component for accelerating research progress, sparking innovation, fostering the renewal of research organizations, and contributing to technology leadership, especially during times of declining support for discovery research in industry.

In the mid-1980s the Board of Directors of Akzo Nobel recognized that the nature of industrial research was changing and that the issues associated with market and technology leadership and capturing emerging technologies were receiving less attention and resources than the shorter-term subjects. To achieve the desired balance of research activities, Akzo established its third corporate research center in Dobbs Ferry, New York. This center was chartered to conduct focused discovery research in core and emerging technologies with a major emphasis on collaborative external research with U.S. universities and national laboratories. The many advantages of an outward vision manifested with external research initiatives are well recognized. One of the key issues is referred to as "backfilling," which addresses the need for discovery research to provide a foundation for R&D that is no longer being conducted by

---

[1]Francis A. Via joined Fairfield Resources International as a senior consultant with more than 30 years of experience managing industrial R&D, intellectual property, and market development at Stauffer Chemical Company, Akzo Nobel, Inc., and GE. He directed Akzo Nobel's Corporate Research-US to capture emerging technologies. Utilizing external cooperative research programs at universities and national laboratories served as the keystone for this corporate research.

industry with the same commitment and fervor as during the major growth years of the 1950s to the 1970s. Another advantage is the technical flexibility that is provided to evaluate emerging technologies while minimizing risk.

The goal of this presentation is to demonstrate the role of external programs in accelerating R&D. Earlier presentations showed the business process of managing new product development by applying a stage-gate system. At Akzo Nobel we employed a typical six-stage system to guide activities from the early idea phase to the final step of commercial production. In reviewing the stage-gate innovation process, it may be unclear how an external program can accelerate innovation. In most cases, an external program is principally associated with the idea stage of discovery research (Stage 1). This activity represents not only the very early project action but also a rather small portion of the resources required for new product/process development.

Many studies have demonstrated that the most critical component of accelerating research and increasing both the efficiency and the return on investment of research is the implementation of an effective decision-making process. The ability to select the "correct" R&D programs has proven to be one of the leading components of a successful R&D organization. This decision-making process involves input from all facets of a commercial organization, including science, economics, marketing, competitive analysis, intellectual property assessment, manufacturing, environment, regulations, customer needs, and public perceptions. The Akzo Nobel experiment has shown that a properly structured external research program can help evaluate and understand the science and thereby assist in the early stages of the decision-making process by providing more state-of-the-art options at minimal risk.

The overwhelming majority of external programs at universities and national laboratories are associated with Stage 1, the idea phase, of a new product development program. Influencing the early idea phase can have an impact on the entire innovation process. Robert G. Cooper of McMaster University is a leading business methods researcher in new product development methods. He offers an online course in the training and application of a stage-gate system. A recent study of new product innovations and developments shows that the idea phase constitutes a rather minor proportion of the resources used over the entire project, less than 1 to 2 percent, and the market analysis was more complete for those programs that succeeded. How can an external university component help this process with such a modest early-stage role?

Those familiar with the stage-gate process recognize that there is a strong overlap between each of the phases. As you proceed to the later stages, especially to piloting the process, there is a growing need to fully understand the science to guide and select continuing approaches. A new product development process was likely justified with a stringently controlled budget to facilitate achieving the targeted return on investment as well as the timing needed to secure a strong or dominant position in the marketplace. Thus, resources to develop the underpinning sciences are relegated to a "nice to have" position but are not seen as a critical step in this business process. More often than not, a working knowledge of the underlying sciences can accelerate new product development, particularly at the pilot and customer evaluation stages. This understanding of the chemistry can be obtained at low risk in the early stages and at modest cost with a university or national laboratory partner.

Speed to market is essential to rapidly capture the high-risk investment for new product development and provide the projected returns. How can external research alliances allow us to more quickly reach this breakeven point?

Two examples have been selected to demonstrate this process. The first involves a selective catalysts development program at Akzo Nobel under collaboration with Mark E. Davis of the California Institute of Technology (Caltech). Catalysts with greater selectivity were needed to improve the performance of a product line of phosphorus-based flame retardants and functional fluids. The Akzo Nobel corporate research team had been working with Mark Davis for more than 4 years on new zeolite synthesis primarily for petroleum refining applications. When this particular challenge arose, the external program at Caltech was modified and rapidly developed leads based on an outstanding knowledge foundation of the technology and an effective teaming, especially effective communications with the industrial partner. Based on external leads, the internal developments at Akzo Nobel were accelerated and led to an independent approach. In less than a year the technology had advanced and was ready for plant trials.

The second example also involves catalysts development. The goal of this project, headed by Leo E. Manzer and Walter Cicha at the DuPont Central Research Station, was charged with developing a new highly selective catalyst for the manufacture of phosgene while reducing the amount of the undesired by-product, carbon tetrachloride. As a result of basic studies by the DuPont catalysts research team, it was recognized that carbon tetrachloride formation arose from chlorination of the carbon catalyst that is used in the commercial process to promote the reaction of carbon monoxide and chlorine.

Manzer's team had to address a challenge involving two incompatible factors. He needed a carbon catalyst that would promote the efficient and selective chlorination of carbon monoxide but that would remain inert for chlorinating the carbon catalyst surface. Through many years of experience the DuPont team has built a knowledge base and scientific network that led to the Boreskov Institute of Catalysis in Novosibirsk, Russia. Alliances with international research facilities are a major trend in external programs. The team at Novosibirsk had developed a unique series of specialty carbon materials and supports. The DuPont team evaluated variations of these specialty carbon materials, and within less than a year and a half the catalysts became operational at the DuPont Deepwater Plant.

Building on core competencies facilitated the success of these two examples for accelerating research through external collaborative programs. Both programs exemplify the findings of a 2-year study conducted by the Industrial Research Institute during the mid-1990s. About 300 members constitute the Industrial Research Institute, from R&D-intensive corporations with chemical and petrochemical companies composing about 30 percent of the membership. During the early 1990s, many members wanted to share best practices for improving the performance of R&D. A 2-year study provided a package of guidelines for procedures, approaches, and recommendations termed the Technology Value Pyramid (see Figure 7.1). One of the study's key recommendations was to identify core competencies and maintain supporting technology platforms. These platforms are composed of the full spectrum of R&D activities, including a working knowledge of internal technology, internal networks, market dynamics, external networks, and leading-edge emerging technology.

Thus, an R&D program can be readily accelerated by tapping into your technology platform and internal network to select focused team members. Commonly, there is little or no physical movement; although a program manager will have authority and accountability for the entire team. With this foundation, a well-designed team can achieve a rapid start and accelerate development.

An additional challenge for R&D management arises after completion of the new product or process development program. Scientists and engineers who have helped navigate the new development effort now emerge as veterans with a valued personal and professional growth experience. With relatively high probability, opportunities become available for these "seasoned" travelers to move to other areas within the corporation. This activity of significant mutual benefit creates a challenge for R&D management. It has become imperative to rebuild and maintain these technology platforms and networks.

I would like to provide several examples of successful external university programs and define approaches to persistent issues such as intellectual property and management strategy. Each year the American Chemical Society (ACS) provides an award for leadership in chemical research and

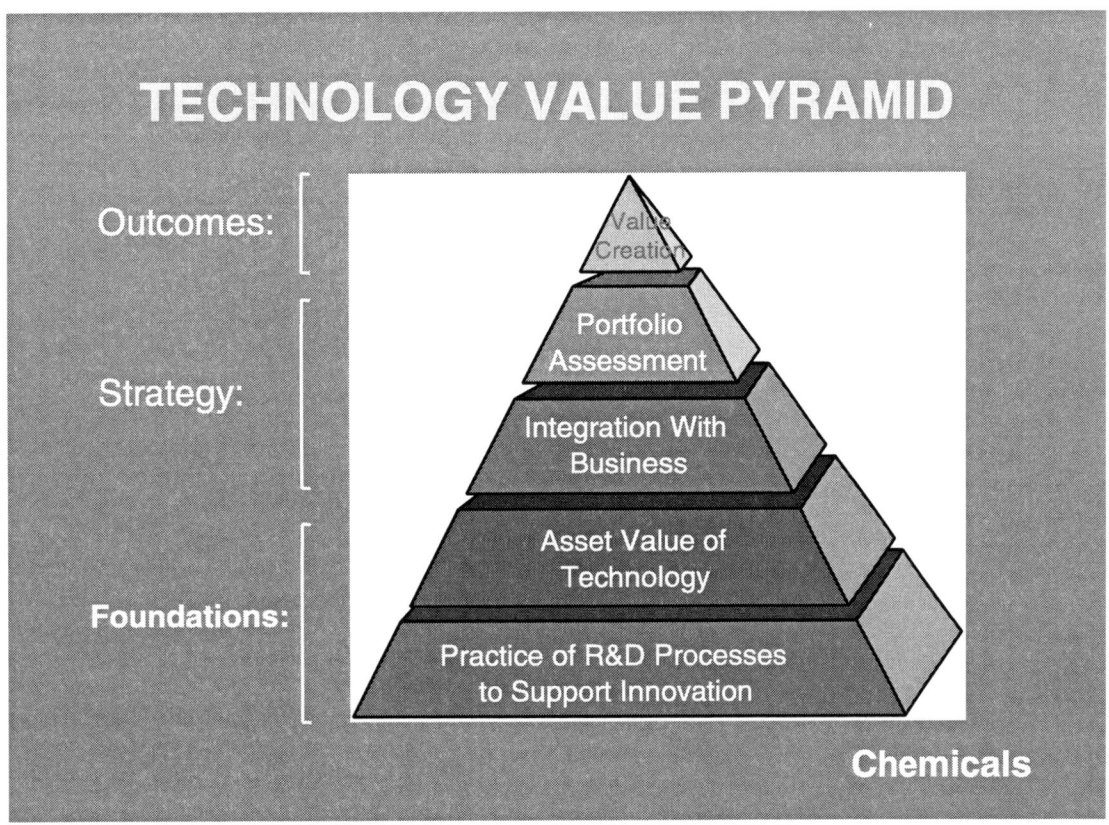

FIGURE 7.1 The Industrial Research Institute's Technology Value Pyramid shows the value of maintaining supporting technology platforms. Reprinted by permission of the Industrial Research Institute, Inc., from "Technology Value Program," 1996, www.iriinc.org/tvp.html.

management—The Earle B. Barnes Award. In 1995 Leo Manzer received the award, which serves as a clear indication that the DuPont team had built a technology platform in catalysis. In 1998 Joseph A. Miller of DuPont was recognized with the Barnes award not only for his management and commitment to corporate research but also for his external vision.

Another example of a successful external research program is the collaboration between Joseph M. DeSimone of the University of North Carolina, Chapel Hill, and North Carolina State University, and DuPont. Dr. DeSimone developed a new solvent system based on carbon dioxide for manufacturing fluorocarbon polymers that avoids the use of chlorofluorocarbon solvents. It required only 4 years from the beginning of the collaborative effort to announced plans for construction of a semiworks plant. This acceleration was achieved after spending nearly 2 years negotiating the research and licensing agreement. A well-designed collaborative program with the right people and targets can overcome obstacles both technical and administrative. How can a university program accelerate a new product or process development project with such common delays for issues such as intellectual property? The successful approach is one of commitment. Both parties must believe the project will be conducted and make plans to initiate research accordingly, protecting all parties in the interim, until an agreement is in place.

This year Kurt W. Swogger of Dow was acknowledged for his vision and leadership in directing the successful Insite Catalysts commercialization program that reportedly was accelerated by a factor of 3

> **BOX 7.1 Key Components for Accelerating Research and Development Through External Programs**
>
> - Objectives and Timing
> - Professors and University Personnel
> - Responsible Industrial Scientists
> - Business and R&D Management Commitment
> - Publication of Results
> - Intellectual Property Ownership Established
> - Relevant and Educational Projects
> - Formal Reviews and Frequent Communications
> - Multiyear Agreement
> - Must Be Flexible

relative to conventional programs. The Dow development team involved five external collaborations with universities for the development of this outstanding new line of olefin polymerization catalysts. Swogger received the Earle B. Barnes Award at the banquet dinner during the spring 2002 ACS meeting. Concentrating, for this discussion, on the role of external projects, it is noteworthy that the Dow award winner is also a champion of such programs. A year earlier Swogger gave the presentation at the spring 2001 ACS meeting on accelerating product development speed via cooperative research with universities. Using ideas from different success stories, a list of key components was assembled for this presentation to outline key criteria for successful external programs to accelerate R&D (see Box 7.1).

I would like to review two of these points: (1) the responsible industrial scientists and (2) the business and R&D management commitment. For these external university programs supported by Akzo Nobel, a secondary but critical goal was to provide opportunities for increased responsibility and personal growth for staff research scientists and engineers. The company also wanted to acknowledge the role of the internal scientists and provide a renewing force for the organization. The position of technical liaison was created, with responsibilities for both the external and the initial internal research programs. To the university the technical liaison represented Akzo Nobel, and to us the technical liaison represented the university. The technical liaison's goal was to utilize the strengths, capabilities, and differences in scientific insight of the university team. Our representatives would visit the university with some frequency in addition to the biennial reviews. For example, in a program at the University of Massachusetts on liquid crystalline polymers, the two technical liaisons would spend half a day with five participating professors and their students every 6 weeks. The most successful forum was the pizza lunch.

For nearly all programs a business commitment was established before starting the external project. Global markets are moving targets, and we have had several external programs fail to maintain the interests of the business group while achieving technical success or making good progress. In some cases, during the R&D program either the market or the market assessment changed or the target business group was divested. Thus, without a potential business commitment or a path to commercialization, an exit strategy for the external program was exercised. In most cases funding for the program was continued at a level to cover commitments and especially to cover the students to graduation.

Metrics for external programs were established that are compatible with the evaluation of an internal corporate research program. A stage-gate system is applied. When an external program reached the level of a business unit funding (Stage 2, internal R&D project), this achievement was considered a success. Overall, about 20 percent of our external programs reached that stage. This number is also a reflection of the risk level and the technical progress of the external program. Other programs have different expected levels of success. The National Institute of Standards and Technology's Advanced Technology Program seeks a higher level of success. On the other hand, a few years ago Texas Instruments funded a series of corporate research projects with the aim of addressing high-risk ideas. These were termed "Wild Hare" projects. When this initiative achieved a 20 percent success level, management questioned whether operations was taking sufficient risk, as a 20 percent success rate was rather high to be addressing programs at the intended risk level.

There are several professional organizations that recognize achievements in external R&D programs. The ACS and the Council for Chemical Research both recognize outstanding accomplishments in this field. I would like to examine two well-publicized success stories for accelerating R&D with an external effort. To develop new chlorofluorocarbon replacements, DuPont formed a team with more than 10 universities and national labs to investigate aspects of these replacement materials and to accelerate the program. External activities complimented internal functions and covered the full spectrum of issues, including structure/property relationships, physical properties, atmospheric behavior, manufacturing methods, and even engineering unit processing studies. This program proved so successful that the U.S. Environmental Protection Agency requested DuPont to delay full replacement of chlorofluorocarbons so that the applications infrastructure could have additional time to adjust to the new replacement materials.

The second example for a successful external project involves the Remediation Technologies Development Forum, which includes members from governmental agencies, seven companies, and volunteering universities. This consortium was established in 1992 by the Environmental Protection Agency to foster collaboration between the public and private sectors. The goals are directed at addressing the technical and regulatory issues for environmental bioaugmentation with a focus on the bioremediation of trichloroethylene. The tasks include developing technology, educating the regulators and the public, and securing approvals from the multiple regulatory agencies—federal, state, and local. The consortium was successfully directed by Dave Ellis of Dupont. The forum was designed to foster public-private partnerships to conduct both laboratory and applied field research to develop, test, and evaluate innovative remediation technologies. A number of test sites have been established, and bioremediation of trichloroethylene was successfully demonstrated. A recent review meeting consisted of several hundred participants, including representatives of federal, state, and local regulatory agencies. Thus, the multitasks of demonstrating technology, establishing credibility, and education were accomplished and accelerated with this external collaborative effort.

As the impact of external R&D on accelerating and facilitating commercialization is assessed, it is essential to classify the nature of the development program. Figure 7.2 shows the characteristic market-technology diagram, where technology (current and new) is defined horizontally and the market (current and new) is shown vertically. Many of the award-winning external programs that were reviewed fall into the new technology for current markets quadrant. This quadrant, popularly referred to as Pasture's quadrant (for the plot of not market versus technology but science versus technology), is the domain of moderate- to high-risk development programs. This area represents a moderate comfort zone for visionary management. How do external programs help the most aggressive quadrant—new technology for new markets? The assessment indicates that external collaborations have a particularly important role to

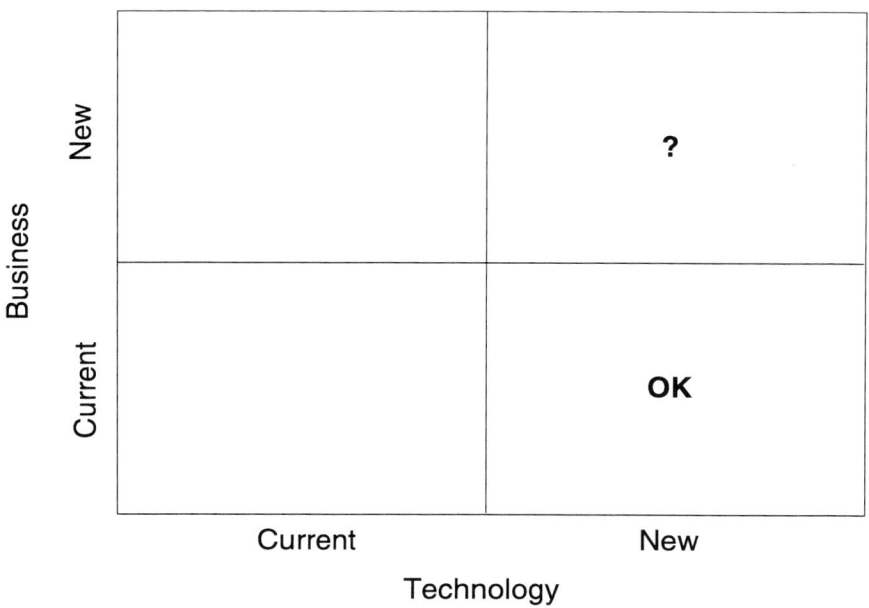

FIGURE 7.2 External programs impact the current business (science)/new technology quadrant.

play in these programs as they can bring to bear different technology issues and perspectives rapidly with reduced risk, timing, and cost.

This next example was chosen because the product, a fast-setting cement, represents a new product for a new market. The term "new market" for this discussion is defined as viewed from the perspective of the potential producer. While highway repair products could be considered a current market for many suppliers, it was a new market for Stauffer Chemical. This project differs from others in that it involved an external collaboration to conduct a market assessment with a business school and not a technical collaboration. As shown, Akzo Nobel worked with a Rutgers business school team that conducted a market assessment study to include a business plan with pricing and a potential customer list. With six students this study was conducted in two semesters and included customer visits in the metropolitan area. The special feature of the product is that the magnesium phosphate cures in less than an hour and a highway can be reopened in less than one work shift, especially between high-traffic "rush hours." This is of value for heavily traveled highways and bridges. Despite the technology and marketing plan, the program failed because reorganizations and divestitures moved the project champions to other businesses. Several planning meetings with the new parties looked encouraging but lacked the staying power for new product development. This example was chosen to emphasize that the commitment of a business team and the role of a champion are two factors necessary to achieve success.

Any discussion on external collaborations with national labs and universities would be incomplete without considering the company's responsibilities to the collaborators and the responsibilities for managing intellectual property. In the Akzo Nobel experience there were four projects for which interest in the business units could not be sustained, and as a consequence, the intellectual property rights were returned to the university. Most of these examples involve new technology for new markets. For these cases involving Northwestern University, the University of Pittsburgh, Rensselaer Polytechnic Institute, and the University of Utah, each university was successful in licensing the technology to third parties. When conducting a successful external collaboration, a window on technology is gained and it is essential to have the commitment to move forward or the opportunity may be lost.

Two other examples of collaboration involve national laboratories where the technology requires multidisciplinary approaches that exceed internal capabilities. Akzo Nobel's expanding pharmaceutical division was interested in exploring new approaches for immunotherapy. A corporate brainstorming session provided the seeds of the project idea based on an emerging technology of radioimmunotherapy. Can the selectivity for cancer therapy be improved by building on the technology to guide an alpha particle to a cancer cell? An alpha emitter provides a steady, linear, high-energy particle. It has a short range and, if guided inside a target cell, can be selectively lethal. All that is needed is an alpha emitter located at the cancer site. The project entails three parts: an alpha emitter, a delivery method, and a way to test it. None of these were in place at Akzo Nobel, so an external collaboration was the preferred choice.

The first action was to select and secure an alpha emitter. For that Oak Ridge National Laboratory was identified as a possible source. Two years later, after implementing agreements and safety procedures, we had secured samples of bismuth 213 with actinium 225 to serve as the alpha source for the program. To develop the radionucleotide delivery system (the cow), the Karlsruhe Atomic Energy Laboratory was contracted. This approach serves as an example of a global collaborative program.

The second project challenge was to bring the alpha emitter to the cancer site. Perlmmune, a subsidiary of Akzo Nobel at the time, produced monoclonal antibodies that could selectively bind to cancer cells of acute myelogenous leukemia. The third challenge, to attach the radionucleotide to the monoclonal antibodies, required the experience and skills in actinide chemistry of Los Alamos National Laboratory. Access was gained via a "work for others" agreement to prepare selected chelating agents. In addition, a tetra acetic acid derivative, obtained in collaboration with Dow, was used for the subsequent studies. David A. Scheinberg, of the Memorial Sloan-Kettering Cancer Center, working with colleagues at Columbia and Cornell Hospital, directed all the pharmacological facets of the program.

Akzo Nobel maintained an equity position as several employees together with a venture capital group formed the spinoff, PharmActinium, Inc. Radioimmunotherapy is a strategy designed to increase the efficacy of native monoclonal antibodies, decrease the toxicity of therapy, and improve the long-term outcome of patients with leukemia. Preliminary trials have demonstrated an extraordinarily high selectivity/kill ratio. With this example, one can see that developing technology with external collaborations provides opportunities that otherwise may not be attainable in a timely and economically palatable fashion.

As a final illustration in the area of new technology and new markets, the involvement of national laboratories is shown to achieve the technical progress and the knowledge integration for accelerating commercialization. As the fad of high-temperature superconductors rose and subsequently dissipated, teams from Argonne, Oak Ridge, and Los Alamos national laboratories together with the University of Wisconsin continued to pursue the science and technology of superconducting wire. As a result of this collaboration, the technology is currently being commercialized with American Superconductor, Inc. About 20 miles of cable are planned for completion by year's end for a utility in the Albany, New York, area.

External research partnerships, properly designed and integrated into internal activities, can accelerate the new product development process. The time from idea stage to commercialization, with the proper commitments, can be reduced substantially, in some cases by 50 to 66 percent. Akzo Nobel has shown that this acceleration can be realized by applying flexibility to achieve a win-win situation for all partners and by implementing the guidelines for external collaborations. Developing an avenue to participate in the research infrastructure at U.S. universities and national laboratories is becoming essential for sustainability and technical leadership in an industrial R&D organization. The advantages of this outward vision are broad based and have been shown to provide a positive impact on industrial research.

In conclusion, the principal foundation for accelerating R&D with external programs requires maintaining key technology platforms; processes for selecting both the programs and the external collaborators; internal and external networks including business teams and scientific leaders; identifying market-driven needs; securing a business commitment; and a renewing investment in the internal R&D team members. I am pleased to acknowledge the many outstanding participants and proponents of this 12-year experiment at Akzo Nobel. It was a privilege to have been a member of the Akzo Nobel team and to have enjoyed the camaraderie and intellectual challenges provided by these leading managers, scientists, and engineers, both internal and external, through over 100 collaborative partnerships at 30 universities and seven national laboratories.

DISCUSSION

J. Stewart Witzeman, Eastman Chemical: One of the problems I have observed with external research programs is that the manager is often fighting internally because the R&D community sees it as competitive and because it is often seen by senior management as one of the first things to cut in tough times. Do you have any best practices on how companies are able to maintain continuity in these sorts of programs?

Francis Via: You have identified a critical issue for external programs that reflects on sustainability and relative value demonstration. Several of the examples given in this review involved programs at Dow and DuPont, which have both successfully addressed this persistent issue. One of the frequently proclaimed advantages of external R&D programs is that there is flexibility with reduced risk. This nature makes them more vulnerable to fluctuations in research funding. This is but one of the many challenges of R&D management in the 21st century. The principal goals of the Industrial Research Institute study leading to the Technology Value Pyramid also address this issue by demonstrating the need to maintain technical leadership in core technology areas. External networks can be developed and maintained in times of financial restrictions with less funding and additional personal involvement.

Andrew Kaldor, ExxonMobil: I have never seen so many examples of entrepreneurial linkages from external relationships. How much overhead, time, and effort are required to manage a successful project?

Francis Via: A genuine commitment is required to conduct a valued external research program. Overall, funding was allocated to the internal R&D team for review guidance and follow-up that was essentially equal to the external funding. The Akzo Nobel corporate group maintained a research team at Dobbs Ferry with full responsibility for conducting the internal research starting after external demonstration of the concept stage. Also this team has responsibility to serve as the technical liaison for at least half the external projects. For the other half of the projects, the business units contributed 10 percent time share the first year and a 50 percent share for the second year to cover their scientist who served as the technical liaison for select programs that had advanced beyond the concept stage. In addition, several of us who were managing the programs were involved full time, screening, reviewing, and identifying projects and gaining input from or selling to the business community.

I would like to comment on intellectual property as that issue was raised earlier. Akzo Nobel had more than 100 contracts with universities. In most of those the intellectual property was assigned to the university. We would write and file the application, covering all costs for patent execution on behalf of the university. In response, Akzo Nobel would obtain the first right of refusal for a royalty-bearing license with limits on royalties defined. The universities maintained the right to license to others in different fields of use.

8

Some New Ideas for Speeding Up the Development of Products from University Research

Kenneth A. Pickar[1]
California Institute of Technology

There is no question from what we have heard that university research is of increasing importance to industry and that this is controversial. Big companies have been neglecting basic research. For example, the number of publications from IBM and Bell Laboratories dropped by a factor of 2 between 1985 and 1995. At this rate it is clear that universities will be of increasing importance to industrial research. Within the industrial portfolio, there are high-risk/high-reward investments and there are low-risk/incremental improvement investments. Universities can, in principle, be leveraged for benefits in both scenarios, but in reality universities are much better suited to the former.

According to Clayton Christensen,[2] businesses are often hit by sneak attacks from new technologies they do not expect. These new technologies threaten to make some company research obsolete. Connections to universities can help industrial scientists be prepared for some of these technological attacks.

I remember the days when industrial scientists felt like they needed to have most research and development under their control. The chemical industry has led the way in showing that need not be the case. Two things that a company needs to do well no matter how big it is are (1) form partnerships and alliances to provide for the research and development they cannot do themselves and (2) be open minded about ideas that come from other sources. Both of these can require working closely (collaborating) with universities.

From the university's perspective, these relationships can be very attractive. First, there is the obvious desire to tap some of the economic value of such partnerships. A cycle exists wherein technology is spun out of universities and makes companies successful. In return, corporate donations and royalties come back to the universities. Licensing revenues can be very substantial. In addition, start-up companies can create very wealthy entrepreneurs who are carefully tracked by the university. Eventually the

---

[1] Kenneth A. Pickar is currently professor of mechanical engineering at the California Institute of Technology and co-principal investigator of the National Science Foundation (NSF)-funded Entrepreneurial Fellowship Program. His previous experience includes positions at Bell Laboratories, GE, and Allied Signal, Inc.

[2] Clayton M. Christensen. 1997. *The Innovator's Dilemma: When New Technologies Cause Great Firms to Fail*. Boston: Harvard Business School Press.

entrepreneur receives a phone call because the university expects some return. The California Institute of Technology (Caltech) and many other institutions have traditionally enjoyed such benefits.

In addition, universities vying for young engineering and scientific faculty require a fairly robust entrepreneurship program. In engineering and applied science, Caltech competes with its peers by showing prospective faculty members examples of past Caltech professors who have successfully spun out new companies. Caltech demonstrates it has an atmosphere that values this activity, at least for professors who are established with tenure!

To meet their goals, universities then want to work with companies both nascent and mature. The result is that industry-university research has grown from about $1.5 billion in 1994 to $2.6 billion in 1999, a 10 percent compound annual growth rate. Industrial support of research at universities as a fraction of total academic effort has grown as well, showing that industry and academia are becoming more important to each other. Despite this encouraging news, the details are not pretty. Even casual observation shows that the amounts of wasted energy, misspent effort, and lost opportunity are very large, especially compared with the normal standards given to managers of industrial enterprises. They are asked to maximize value, minimize defects, and increase productivity from all of their activities.

Thus, a target-rich environment exists for improvement. The root cause for the dysfunction is the cultural impedance mismatch between academia and industry. From the industrial perspective the university often shows amazing naïveté. The average professor will assume that he has created a great technology. He will then grossly underestimate the amount of effort required to commercialize this technology, with major challenges like distribution, financing, and marketing undervalued.

The notion that it is possible to sprinkle a relatively small amount of money and effort on a great technology to create a commercialized product is, of course, completely false. Our students of entrepreneurship develop their business plans based on a developing technology. It can take them 5 or 6 months of commercialization activity before they realize some of the profound business difficulties associated with their activity.

There is also a suspicion of entrepreneurs based on the cultural misunderstanding of business. Universities fear that they are going to be exploited, that the business people they deal with are fundamentally dishonest. Academics fear that business partners will spend too much on the business and not share enough with their university partners. First-time entrepreneurs can be so suspicious of their corporate partner that they will hold back important information. They are afraid that their business partner will expropriate their product because academics feel it is trivially easy to steal intellectual property just as it is trivially easy to commercialize the resulting technology. Although business schools associated with many universities could help here, with notable exceptions there is an amazing lack of academic interest in the study of technology management and commercialization. Thus, it can be an uphill fight to gain academic recognition in business schools for technology commercialization.

Another barrier to the industry-academia relationship is the "if it's not broken, why fix it?" mentality. Presently, many universities are quite well supported by government research funding. There is a fear that stronger ties to industry would compromise the current direction of university research and, worse, could even compromise scholarly pursuits.

Thus, it is no surprise that commercialization results are far from optimal. University research that could be converted to useful products is done not at all or done imperfectly. In a process sense, the latter stages of the development process can have very low first-pass yield. Consider as well the education process. Due to the lack of experiential understanding of the reality of the business world, students can emerge from the process essentially poorly educated. For example, as a result, those entrepreneurial graduates who start their own companies are prone to make mistakes—not just unavoidable mistakes but mistakes that could have been avoided.

To show how we have addressed these issues at Caltech in the entrepreneurial arena, I will discuss two areas. One is the Caltech Technology Transfer Office and the other is the Entrepreneurial Fellows Program, which I have been working on for the past year and a half.

The Caltech Technology Transfer Office has issued approximately 40 to 50 patent licenses and options each year. That makes it approximately number three in the nation, which is remarkable for a comparatively small institution. In short, the Technology Transfer Office succeeded because it convinced the faculty that they were there to serve the professors, not the administration or the venture capital community.

This is not to say that the Technology Transfer Office did not secure top-level administration support and venture capital support; they did. But the faculty was central. They identified the minority who were the key faculty interested in commercialization, and they worked on building confidence. They worked hard to get some early "good examples," key to change a culture. In particular, they were very aggressive in filing invention applications. They understood that without intellectual property protection it is very hard to raise money and that obtaining that protection is very expensive. Thus, the Technology Transfer Office files provisional applications for all new invention disclosures—120 patents were issued in 2000, a very high number for an institution with only 275 faculty members.

To encourage new commercialization ideas, the Technology Transfer Office created a "grubstake" program, a small venture fund supported by alumni and trustees. Up to eight projects are supported each year, at $30,000 to $50,000 per project. The concept is to develop technology that is not "ready for prime time"—not ready to be commercialized. For example, the money can support the development of an experimental prototype. There have been 30 awards since 1995; 10 have resulted in licensing agreements to companies and seven form the bases for new start-ups. To emphasize the faculty-centered nature of the program, a Caltech faculty member has to be involved in the project.

Faculty members typically do not take the lead in this program or start their own companies. Their graduate students are the ones who take the lead, often with great passion, commitment, and a desire to change the world. The students work with experts to develop a preliminary business marketing plan. The Office of Technology Transfer then arranges for various service providers to assist, including law firms and certified public accountants.

Although the professor typically creates the invention and writes the patent, Caltech owns it. However, if the professor wants to practice this technology exclusively commercially for a period of time, he or she can take out an option on the technology for a relatively nominal sum. The professor must commit to using the patented technology and make it successful within a 12-month period. If this is not done within 12 months, the option reverts back to the university and exclusivity is lost. If, however, the option is exercised, a company is formed, and Series A money is raised, Caltech will take a small percentage share of that company in return for granting an exclusive license.

Thus, the entrepreneur does not have to pay out any money to Caltech but rather grants Caltech several percentage points share of the company. In addition to this investment interest in the company, Caltech will continue to support the company with various services. The professor who founded the company is also motivated to support the nascent company with a continuing stream of ideas and new inventions.

A mutually beneficial "partnership" was formed as the Technology Transfer Office demonstrated it could move with simplicity and speed. It was able to achieve high commercialization productivity, and I would argue that this was done without interfering with the academic mission.

I would like to talk now about the Entrepreneurial Fellows Program. It is a joint effort of Caltech and the Art Center (College of Design), funded initially by the NSF. Both of the participating institutions are small and without strong business strengths. However, many of the students at both schools are interested in entrepreneurship. The purpose of the Entrepreneurial Fellows Program is to educate students with an entrepreneurial frame of mind, not, as its prime purpose, to create new companies.

There are a few career choices available for students who wish to become entrepreneurs. One is to enter directly into an entrepreneurial enterprise, make the normal mistakes, and then try again, learning from mistakes. An alternative is to advance to an M.B.A. degree, usually entered after a few years of employment. Another option, one that I often recommend to my students, is to join a company that has a successful business training program and good prospects. For the entrepreneur it can be effective to join a first-rate company, learn, and then invest that learning by starting an entrepreneurial enterprise.

For Caltech students we decided to create another path to redress this gap between what the students learn in academia and what is required to start a company. We developed a relatively short-term but very intense business training program that would prepare them for the entrepreneurial world.

The first Entrepreneurial Fellows Program lasted 9 months, during which students behaved as if they were part of a new company. The second iteration will be 6 months because of program improvements made based on the first program. Participants have to be recent graduates with a B.S., M.S., or Ph.D.

Universities train students to do independent research, but this is irrelevant in a company environment. Students who are trained in a very competitive environment often do not have good cooperative experience, so our program is team based. Each team has its own mentor to assist it and aid the learning process. The fellows receive a stipend, and business expenses are covered as well.

Our instructional staff are practitioners rather than professors. We have been very successful in obtaining the services of first-rate experienced people, often consultants, who come multiple times during the period of instruction to teach, spending additional time coaching our students. We use local angel investing groups (for example, Tech Coast Angels) and venture capitalists as well.

We use a mixture of traditional and nontraditional teaching methodologies. We integrate traditional Caltech courses in product design and accounting into the program. Other teachings are typically divided into 1- to 2-day modules so that new knowledge can be immediately put into practice in a "learn/do" fashion. The program also involves a number of "charettes," rigorous 2-day exercises during which students are given a problem, brainstorm concepts, and are required to present to an audience. The intensity and the team atmosphere help spur unusual ideas. Students also observe Tech Coast Angels meetings, watching and critiquing as entrepreneurs make business presentations. This affords the students an unusual opportunity to understand the standard for getting funded today.

The curriculum contains all the obvious things you might expect. We begin with marketing and team building. The first weeks are spent calling potential customers and sizing market needs. We deliberately avoid technology development in the beginning to emphasize the importance of marketing in developing a product plan. We understand that students typically will work on technology because that is their comfort zone. We want to challenge them to think beyond that.

We have finished with the first group of fellows who graduated at the end of March. The second group has been selected. The program is relatively expensive because we pay our entrepreneurs as fellows. To make the program more affordable, we have reduced the stipend for the second group and reduced the program duration from 9 to 6 months.

All involved in the program enjoy working with these young people. We know that among them are going to be one or two who will make a big difference to society, perhaps changing the world as we know it. There is a lesson to be learned from this. Some of the students will present ideas that you know wouldn't survive investor scrutiny. There is an issue of balancing expectations against reality. To resolve this dichotomy, we are all rigorous in defining our program as an educational experience. We will measure ourselves by the quality of the education provided, not by the number of businesses spun out. Our students will then have a learning experience that they can apply to their present ideas or to develop new ideas.

Finally, what can be done to encourage industry-academia collaborations? I have some experience in good business practices based on my background in industry at companies such as AT&T, General Electric, and AlliedSignal. I am now at a university, so I have added that perspective, too. I agree with the notion of simplicity that Jack Welch and the previous speaker, Jim Heath of UCLA, both espoused. Universities need to be easier to do business with. Clearly, there must be some flexibility in negotiation, and negotiation should not take months.

Companies, too, must be simpler with a more standardized interface. They need to be proactive and flexible, involved with universities early and continuously. There are many poorly packaged commercialization opportunities as presented by the universities. What's more, they don't know what they don't know. What does a professor typically know about the various applications of his technology?

Companies are in the driver's seat. Why shouldn't they initiate the process? One thought is for a company to package its needs and then do a little shopping among universities, that is, compete individual universities. This is possible if a company has a well-defined vision of its technology needs and knows what problems need to be solved to achieve that vision. The entire search process can be much better defined and a long-range relationship decision made based on the quality of the professors, the professors' desire to work with you, how easy it is to work with that university, and so on.

This is intended as an entree and as a focusing exercise to prove to busy people that there is mutual interest in cooperating. It is also useful to concentrate on one or two good examples and demonstrate success. Internally to the company, I have learned from experience that someone in the company who really cares about the project is required. For example, it is clearly a waste to fund a specific university because one of our employees was an alumnus but had no technical or business interest in the results. After writing the check, we had no other close ongoing relationship with the research until it was time for renewal.

However, companies should exploit their excellent networks of employee-alumni who know both the company and the university. They can serve as the eyes and ears to bring together company needs and university technologies.

It is well known that the best way to transfer technology within a company is to move people. For example, researchers are often employed to help scale up a research process to production. In a similar fashion companies can work with graduate students, giving them the understanding that if everything works out well they could be hired to help transfer the technology. In addition, professors could be hired for sabbaticals in corporate research laboratories. Finally, projects could be considered for joint funding. A funding organization like the National Institute of Standards and Technology, NSF, or the Defense Advanced Research Projects Agency could drive a program that helps bring industry and academia together. This is a win for everybody.

DISCUSSION

James R. Heath, University of California, Los Angeles: Once your students have completed this program, what do they go on to do?

Kenneth A. Pickar: We have only completed 1 year. There were nine students. Some have joined other companies, some have continued to pursue their ideas, and some are going to graduate school. My contention is that those who go to work for companies will be better having gone through this because they will know something about business.

David E. Nikles, University of Alabama: When you have professors getting involved in being entrepreneurs and spinning off companies, you said that it doesn't impact the academic mission. Is that really true? How much intellectual effort are they putting into these companies, pursuing NSF support while keeping graduate students busy and teaching classes?

Kenneth A. Pickar: There have been cases of abuse. There is at least one case I know where a professor was spending more time running his company than performing his duties. That was a tenured professor who is no longer at Caltech. I believe he was asked to make a choice.

There may be inappropriate decisions being made now as we speak, but if you look at any measure of Caltech's productivity in the sciences, its government funding, or its publications, there are no signs that somehow position and privilege are being abused.

A question was raised earlier about whether or not the students who work for these start-up companies are being exploited. I think there is a greater potential for abuse here. It is very difficult to get a job, especially in academia, without your professor's recommendation. There is a correspondingly strong desire not to do anything that makes your professor angry. In some cases there have been students who were afraid to ask about starting a company because they were afraid their professors would think less of them. In other cases, professors were managing students who had developed their own company, treating them like graduate students even though they were postdoctoral scholars. I do not know how to quantify the extent of this problem, but I think it could be of concern.

Ned D. Heindel, Lehigh University: The Chemical Sciences Roundtable met here a couple of weeks ago, and we heard from several of our friends from the National Institutes of Health and research corporations that big pharmaceutical firms have very recently changed their position on the patent story. You indicated on one of your slides that you have an aggressive tech transfer office that would seek to obtain those patents, take care of the cost, and protect the technology early on. What we have been hearing from big pharma is that if you have a patent, don't come to us. There isn't enough time running on it for us to take it to a commercial product. So if you patent a pharmaceutically related product, the industry is not interested.

Kenneth A. Pickar: That concept may relate to drug discovery, which is an exceedingly long product development process. Those companies that I am aware of sell enablers to drug discovery. These inventions are patented and venture capitalists are funding them. If you build a successful product based on an enabling technology, big pharmaceutical companies are candidates to buy you, representing your liquidity event.

Nancy L. Parenteau, Organogenesis, Inc.: You described a wonderful program to educate students on entrepreneurship, but the founders of these companies are in reality the faculty members. They are full-time professors with one or both feet in their academic labs, and in many cases they are going to have a very influential role in the spin-off company. As you've mentioned, they have sometimes too big an influence on the students that then go to the company.

Therefore, can you envision a training program in entrepreneurship whereby the influential faculty can benefit? I find that in the biotech field there are people who really are in need of such a program. I think it is fabulous that the young people are assisting the faculty, but they are not going to be the chief financial officers, presidents, and chief executive officers. They are not going to be the heads of those companies. They will be very valuable employees and will climb the corporate ladder, but what about those influential faculty members?

Kenneth A. Pickar: I agree. There is a need. There are programs on campuses for entrepreneurial students but much less so for entrepreneurial professors. One exception I attended recently was at the UCLA medical school where, under the aegis of the UCLA Technology Transfer Office, the Tech Coast Angels, a Southern California investment group, put on a seminar for about 40 faculty on the capital acquisition process.

However, it would be safe to say that based on my experience many professors lose interest in a company once it is started. They will get some stock and that is it. The professors see their students as having opportunities to venture into business, but it is not something that they themselves are obsessed with.

Mary L. Mandich, Bell Labs: Does Caltech have a business school?

Kenneth A. Pickar: I am the business school for Caltech. Caltech does not have a business school!

Mary L. Mandich: Exactly. It strikes me that what you are actually doing is starting up a grassroots business school. Is there a need in the larger business school community for the type of program or course that you have here?

Kenneth A. Pickar: That is a great question. Stanford has a fine business curriculum that was created by the engineering faculty because they couldn't get enough attention to their problems in the traditional business school program. So they have the functional equivalent of a business school, and they have developed many companies from that engineering-based business school. I love that model. It is entrepreneurship at work in an educational environment. I will put in another plug for a program that is led from the University of Southern California also on an NSF grant. N2Tech is an entrepreneurial network of universities all over America, including the University of Arkansas, University of Pittsburgh, Cornell University, and others, whereby teams are formed and ideas are shared to benefit us all. For example, Rose-Hulman Institute of Technology is a great school, but it is not located in the center of technology in Indiana. Nevertheless, Rose-Hulman is able to get access to lawyers and venture capitalists. There are other experiments going on all over the country, but there is still a huge amount of opportunity.

Joseph S. Francisco, Purdue University: I would like to add that the chemistry department at Purdue is considering partnering with the business school to supplement current courses and provide additional courses in entrepreneurship and business education without having students enroll in an additional 2-year program. Our research department has discovered a need and considers this an investment because it is good for the graduate students.

Kenneth A. Pickar: The University of California, Berkeley, also has a Technology Management Certificate Program where students take classes in both the business and the engineering schools.

Joseph S. Francisco: I think this is an increasing trend.

Kenneth A. Pickar: There is no question. In October at Stanford University there is a roundtable on entrepreneurial education. This year the attendance will be limited because it has become so popular since its inception 3 or 4 years ago.

Joseph S. Francisco: The University of Pennsylvania also has a program.

Kenneth A. Pickar: Yes, and of course they have one at the Massachusetts Institute of Technology's (MIT) Sloan School of Management as well.

Mary L. Mandich: This is a natural outlet for graduate students who want to go on in business as opposed to going to Wall Street.

Michael Schrage, MIT: We had exactly these kinds of problems at MIT. However, unlike the situation at Stanford where the engineers ultimately became disgusted with the business school and set up their own shadow business school, things have been a little more cooperative at MIT within Sloan. This is reflective of a very important trend that merits special attention because the business schools are dealing with the same level of theory divorced from contact with the real world that often times chemical engineering and chemistry are accused of. As a result, engineering departments create their own business schools simply because of what the business schools have become. It is difficult to imagine which business school faculty you would go to for help with an entrepreneurial program.

There is a real opportunity for chemistry departments, chemical engineering departments, and also at MIT again because of the material sciences. There isn't a real activity there, and it has to move away from a consulting model and more toward serving on advisory boards. How many people here serve on science advisory boards at for-profit companies as opposed to management? One would think for a crowd like this one would get a disproportionately high number. To be on corporate advisory boards might be very interesting for deans and faculty.

Of course, BP has a very good training program. Exxon has some excellent programs, and people then get good exposure to business issues even though they are brought in primarily because of their technical expertise. This becomes a nonthreatening way of dealing with some of these problems due to lack of business skills.

Kenneth A. Pickar: When I was at GE and wanted to evaluate a program, I would call in the expert professors from MIT and Caltech, who would advise me on whether these programs were really what my scientists told me they were. I worked with about 20 professors at various times over the years. We have heard it from many people today—big companies are much better integrated with universities than they ever have been in the past.

Ned D. Heindel, Lehigh University: I happen to have a reprint here from *Technology Review* for September 2001[3] which lists the institutions with the number of patents obtained. Caltech is right up there—third or fourth in the nation. Then institutions are listed based on economic return, and Caltech is down around number 20. I quote that "despite the glamour in entrepreneurship, the big money for a university usually comes from patents licensed to large, established companies—not start-ups." You indicated your administration is extraordinarily supportive of programs like that. Why?

Kenneth A. Pickar: I could say that the students who wanted to start companies and the professors wanting to start companies have outweighed those relations we have had with companies that wanted to license technologies, but there is huge opportunity in both.

Ned D. Heindel: It is curious that the university administration is so supportive of cash and assistance and creating start-ups when apparently the better approach from the provost's point of view would be to return the bottom line by licensing the big companies.

---

[3]H. Brody. 2001. The TR university research scorecard. *Technology Review* 104:81-83.

Kenneth A. Pickar: That is a good point.

Mary L. Good, University of Arkansas, Little Rock: The issue is that there are always those small companies. If you look at the ones that have gotten started, it is not that they give money back, but when the university starts a capital campaign, it knows where to go to get the money. Being on the Rensselaer Board of Trustees I can explain that to you in detail.

Peter A. Koen, Stevens Institute of Technology: I wanted to comment about a cultural issue within the engineering department. I am in the technology management department at Stevens and teach a course to executives from companies on how to be a corporate entrepreneur. It is a 14-week course for two to three hours a week. At the end of that time, half of the groups get funding internally in their company anywhere between $100,000 and $500,000. In fact, 2 years ago one group actually got a commitment of a billion dollars at the end of 14 weeks.

The typical student is 35 years old and 10 years postgraduate. The dean of the engineering school called me over and said, "We would like you to teach that same course to the Ph.D. students to help them be more entrepreneurial." I replied, "Great, I would love to do it," and then he said, "But we can't give you 2 hours per week for 14 weeks. Could you teach it an hour per week in 5 weeks?" I asked, "Do you understand that there are many marketing issues such as understanding championing and putting together a business plan?" and he said, "Yes, yes, I understand that, but can't you do it in 5 weeks and only an hour a week?" I told him it would not work, and he did not quite understand that. I didn't know what to say. Starting my career as an engineer I didn't understand the need for management until I had to deal with real management issues. Thus, there are real cultural issues that I don't know how to get around.

Kenneth A. Pickar: I suggest choosing only one topic such as marketing or finance. Just do one, do the best you can, and make sure there is plenty of homework they have to bring in to make the best use of that time. Make it a tutorial where you spend a half hour in the tutorial, but let them do all the work. If they get captured by the marketing, say, "I am doing finance next time."

9

From Molecules to Materials to Market: A Rational Framework for Products Formulation and Design

Venkat Venkatasubramanian[1]
Purdue University

I will address the early part of the innovation cycle—the discovery in the early stages of a project where the design space is explored for improvements on formulation based on the original idea and how that explosive research base is managed using modeling and knowledge-based techniques. I will start with some background on product formulation and design, because that phrase means different things to different people. Then, with the aid of some industrial design case studies, I will argue the need for a rational automated framework through examples of different design problems. Finally, I will summarize the lessons we have learned.

When I speak of product formulation and design, I refer to the systematic identification of the molecular structure or material formulation that would meet a specifically defined need. In other words, you know what you want, but you don't know what structure or formulation will take you there. This fairly broad definition is applicable to a wide variety of situations. For example, engineering materials, polymer composites, catalysts and fuel additives, agrochemicals, and pharmaceutical problems all fit into this framework.

I have three examples and will start with the design of fuel additives, which is a somewhat simpler molecular structure design problem. Then I will mention rubber compounds formulation, a material design issue. Finally, I will mention some ongoing effort in the area of catalysts.

Overall, a company is interested in the move from molecules to materials, the use of those materials in components, and the integration of the components into a final product. Typically, chemists and chemical engineers work at the early stages of the chain of making materials; then the materials are tossed over a proverbial brick wall where mechanical and industrial engineers make components and particular products.

The chemists and chemical engineers on one side of the wall do not have much interest in or understanding of the constraints of manufacturing and design, and the mechanical and industrial

---

[1] Venkat Venkatasubramanian is a professor of chemical engineering at Purdue University. He has been a consultant to several major global corporations and institutions, such as Air Products, Arthur D. Little, Amoco, Caterpillar, DowAgro Sciences, Exxon, Lubrizol, United Nations (UNIDO and UNDP), Indian Oil, ICI (U.K.), Nova Chemicals, and G.D. Searle.

engineers on the other side of the wall certainly do not worry about the more basic research issues. Design choices and manufacturing decisions are made subject to some constraints, some of which could have been avoided, if the decision makers were aware of what those constraints are. There is inefficiency in the overall design process in going from molecules to engineering materials to markets.

The first example is a case study in the molecular design of fuel additives. Through the combustion process, undesirable largely carbon-based molecular fragments are created that are deposited on the surface of the intake valve. Over time these deposits accumulate and eventually inhibit the proper opening and closing of the valve, resulting in suboptimal combustion and noxious gas releases.

Therefore, the U.S. Environmental Protection Agency (EPA) has mandated a test. Before a fuel can be sold, it needs to be tested in a standardized engine. Previously the engine was from BMW, but now EPA is using American models. The BMW engine is run for 10,000 miles; then it is taken apart. Deposits on the valve are measured to determine the intake valve deposit (IVD). The IVD needed to be less than 100 milligrams before the fuel could be sold. There is a whole market for fuel additives, which trap undesirable molecular fragments and prevent them from depositing on the surface of the valve. The problem is how to design these fuel additives to minimize the IVD and ensure it is 100 milligrams or less.

To exhaustively test all possibilities is very expensive, because for every 10,000-mile test the engine must be disassembled to measure the IVD. Every single data point costs about $8,000 to $10,000 and a considerable amount of time. Therefore, we were asked to develop a model-based approach to this problem in 1995.

A second problem involves Caterpillar, which sells earth-moving equipment. This equipment is largely made of metal—in fact, 99 percent of Caterpillar's machinery is iron—however, there are more than 1,000 rubber components in these machines, including tires, hoses, engine mount gaskets, and other parts. This equipment is used in oil wells in Siberia in the winter as well as oil fields in Kuwait in the summer. When this kind of equipment fails, it is usually the rubber components that give way because of the extreme operating conditions they face; hence a multimillion-dollar machine is sitting idle because a $1,000 rubber component has failed. This is a major product, liability, and warranty headache for Caterpillar.

Component failure is so crucial that Caterpillar does not trust any other company to make these rubber products—not even Goodyear or Firestone. Caterpillar makes its own rubber component formulations. Rubber component failure is a multilevel issue: performance depends on the rubber parts, which depend on the rubber component-based materials. This, in turn, depends on the failure mechanics properties of these materials, which are affected by rubber curing chemistry. In the end, the design- and manufacturing-related issues depend on quantum chemistry of sulfur links. This is another problem in which the transformation process goes from molecules to materials to market and has the proverbial brick wall in between.

The challenges and designs we typically see involve very complex chemistry and highly nonlinear systems and processes. There is usually some understanding of first principles and fundamental physics and chemistry but not enough to complete the parts design or the molecular design.

One other problem is combinatorially large search spaces. There are 100 million potential candidates for rubber components formulation that are possible. Some other examples we have worked at Purdue have involved 10^{20} to 10^{30} different molecules. Another issue is that typically there are limited and uncertain data. Most often, combinatorial chemistry approaches do not succeed in these cases because obtaining the results is time and labor intensive. Fuel additives design, for example, requires dismantling the engine for every test of a new formula.

What does this mean with respect to design? The traditional approach has been to give a senior experienced engineer or scientist some design objectives and have that person hypothesize a particular molecule or formulation. Guesswork, intuition, and experience are used when the new molecule is synthesized in the laboratory.

After the molecule or formulation is made, it is evaluated to see whether it meets the objectives. If the process is not successful, it begins again. This typical guess-and-test methodology yields a very long and expensive cycle (see Figure 9.1). Clearly there is a need for a more rational approach, which will remove some of the guess-and-test elements. These problems are so complex that guessing and testing cannot be completely eliminated, but the development of a system that can increase efficiency can help.

For an automated approach, two problems need to be solved. The first is how to predict the macroscopic properties given the structure or the formulation. The second is how to identify a structure based on a given desired set of properties.

There are three options for modeling choices. Fundamental models depicting the chemistry and physics of the problem can be used to predict material properties, although this type of model is uncommon. A second option is to depend on the experience of formulation scientists by using a rule-based model such as qualitative reasoning or expert systems, as with the guess-and-test approach. The final method is a data-driven approach, where data are used to make correlations, largely ignoring the physics and chemistry.

Historically these problems have been examined using one single approach, while we feel a combination of all three is needed. Understanding the physics and chemistry can provide a base, expertise can guide the search, and data can refine it. The question is how to develop a hybrid framework that mixes all three.

We have used physics and chemistry (including quantum mechanics) to build a primary model. From this and the experience base, some intermediate-level structural descriptors have been developed and mapped to the performance using data-driven techniques, whether they are statistics or neural networks. In this way the model can be validated. With the hybrid model we can predict the properties given a structure, and for the inverse problem we can search through design space for properties using a genetic algorithm and obtain the molecular structure or the rubber components formulation.

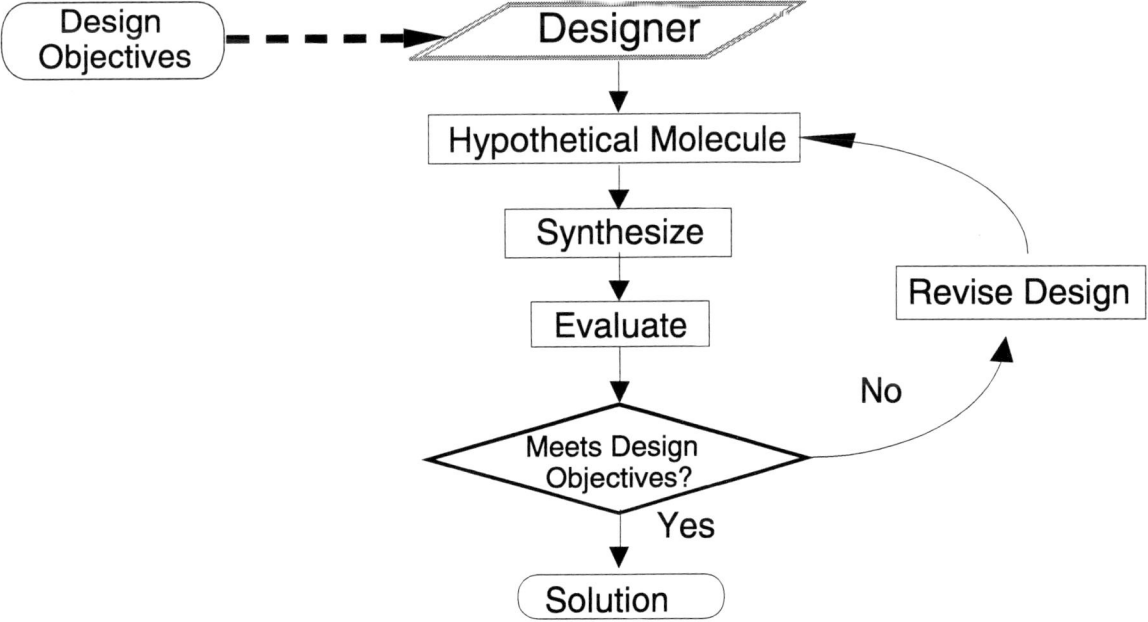

FIGURE 9.1 The traditional design method is a lengthy and expensive process.

Expecting the EPA to lower the IVD value in the future, we were asked by Lubrizol to design a fuel additive for an intake valve deposit of 10 milligrams. We used a genetic algorithm in the hybrid model (see Figure 9.2) to predict the properties of some designed molecules.

One structure we discovered that came close to meeting our needs (99.3 percent fitness, 12 milligrams IVD) had been already discovered by the Lubrizol scientists through their intuitive guess-and-test approach. However, the hybrid model discovered two other better structures. The best of the three had completely novel chemistry and was a combination of molecules that had never been thought of. The hybrid model used "out of the box thinking" that opened up possibilities of new chemistry for generating leads in a much shorter time frame.

I would like to return to the rubber components situation again. Many things go into rubber, including activators, sulfurs, retarders, accelerators, and so on. A very interesting and complex set of approximately 820 reactions occur that result in curing. Current models cannot handle so much information; we need a more complex modeling environment for this type of situation.

Of the top three rubber formulations the model designed, one had already been found by the formulators at Caterpillar. It meets the design criteria, but it degrades much more quickly than desired. The two other formulations have much better degradation kinetics. The model found better formulations in a matter of hours when it would have typically taken 2 to 4 weeks. This is a 10- to 50-fold reduction in the time it takes to design a better formulation.

The third area that we started working on about a year ago in collaboration with ExxonMobil is the design of catalysts. This is a different type of product design from the other two examples because here

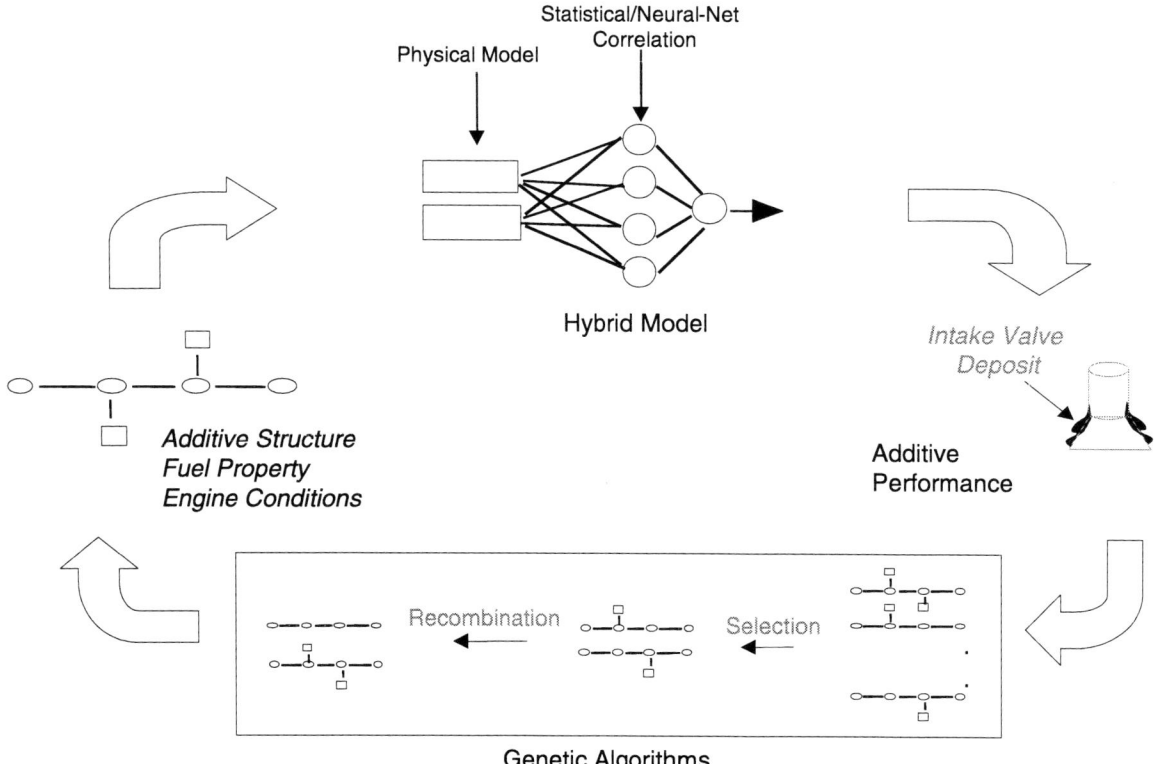

FIGURE 9.2 Computer-aided design, in this case of fuel additives, makes the design cycle much shorter and more efficient by more narrowly directing the molecular search.

combinatorial chemistry can have an impact. In the traditional approach of product development through experimentation, measurements were made one at a time, so there was time to think about how to develop the models, the hypothesis, the mechanisms, and the candidates to fix the data that you are getting. The current approach obtains a lot of data, but the thinking process—the model development process—is still slow and methodical. To get the most out of combinatorial chemistry, the ability to extract knowledge, not just data, is needed. That knowledge can lead you to understanding of the process.

Previously, experimental chemistry and modeling were in sync. They were like a horse and buggy on a dirt road. Now, combinatorial chemistry has provided the experiments with a Ferrari, but it can't be driven at 200 mph because we still have the dirt roads that can accommodate only 20 mph traffic. Modeling capabilities are not on par with experimental capabilities. Initially, there will be successes that are obvious and easy to find. Once you have exhausted those, the next solutions will require true knowledge and understanding. Product development will be limited by the dirt road. At Purdue we believe that the interstate is needed, and we see ourselves developing that infrastructure—the modeling infrastructure—to handle the combinatorial chemistry data explosion. We need a modeling superhighway to get the most out of combinatorial chemistry.

What do we mean by a modeling infrastructure? Given the situation in which some high-level chemistry is hypothesized, too much data exist for one scientist to analyze. For example, the 820 reactions of the rubber components formulation involve over 100 chemical species. It is impossible for one person to write and solve over 100 coupled differential equations without making mistakes. It would take nearly 3 months to explore one scenario. So we have built an environment where the scenario is specified and the information is automatically translated into equations, the parameters are optimized, and the modeled results are compared with data. This way a scenario can be analyzed in a few hours instead of 3 months. This is the type of modeling highway that can get the most out of combinatorial chemistry situations.

For catalysis development our modeling highway can be explored to see what scenarios fit the data. Predictions of new catalysts are based on that information; this may give the desired performance for those catalysts, or at the very least there are new data that could improve the model. The data are useful to revise the model, whether they indicate negative or positive results. Guided experimental design will indicate what part of combinatorial space should be explored.

The guess-and-test approach to product design and formulation is too slow. Combinatorial chemistry can yield much data, but I believe these are not data that we want. Knowledge and understanding are what we desire, and they provide motivation for a model-based framework. I have tried to illustrate these concepts with three examples of actual industrial design problems that we have worked on at Purdue: fuel additives, rubber components formulation, and the design of catalysts. So far our results have been able to reduce the design time or the formulation time. More importantly, modeling has also led to better formulation, new chemistry, and the understanding of driving forces for all of these problems. Nevertheless, we have just scratched the surface of this complex problem domain.

DISCUSSION

Richard A. Sachleben, U.S. House of Representatives: You have shown us three examples of how you used computational methods to address real-world problems. To follow your highway analogy, is it going to require building a new highway every time you have a target or are you aiming toward a generic modeling system that you can utilize regardless of the target? Designing a new modeling system every time you have a new problem to solve is too difficult.

Venkat Venkatasubramanian: That is a very good point, and as you may suspect, we do have a general framework. While the overall tools and the software architecture are the same for each application, a certain amount of customization will be required. We hope to decrease the model customization time from months to weeks, but there will always be some time needed because the chemistry of each problem is different.

Hans Thomann, ExxonMobil: I have two questions. The first regards multiscale modeling. As you probably know, there have recently been tremendous advances in computational metallurgy, particularly in linking length and timescales through parameter passing or imbedding. I didn't hear you mention that. Are you using these approaches?

Venkat Venkatasubramanian: Yes. In the Caterpillar work we use these methods because we are going through different scale levels. The details of how we do it are somewhat different from some of the work that has been done in the computational metallurgy, but the spirit of it is the same.

Hans Thomann: For my second question, I am curious to know a little more about the tradeoff between the different components of the hybrid when you start out with a constitutive relationship and then use some expert knowledge. There must be some weighting value because there is a tradeoff between the time you allocate both the computation and simply putting a weighting factor on the experts' opinion. How do you handle that?

Venkat Venkatasubramanian: Due to time constraints, I didn't get a chance to talk about that. This is an interactive framework. It is not a one-task deal. You are interacting and guiding the search at any given time, and based on your intuition, you can direct the search and change the weights and so on.

Participant: [Comment off microphone]

Venkat Venkatasubramanian: No. There are two ways we handle the knowledge-based guidance. It can be done in real time with the modeler going back and forth between iterations and then guiding the iterations either in the forward model development method, for which the modeler proposes different scenarios, or in the inverse model method, for which the algorithm does the search. The modeler can actually stop the search and force it to go some other direction, based on intuition and experience in how the molecular structure evolves.

The second way to handle knowledge-based guidance is imbedded in both the forward and the inverse model development but typically has fuzzy logic parameters. Now, there you do need some tuning. But you are not bound by that mix alone. That is why it works to first sit down and interact and then overrule where the direction the system is moving.

Participant: [Comment off microphone]

Venkat Venkatasubramanian: Limited visualization. Right now we don't have these fairly sophisticated visualization tools that have come up. At this point we can watch how well we are attaining various properties as molecular evolution proceeds, in addition to observing how we are approaching acceptable performance levels. In some other cases we are undershooting or overshooting our goals, but we can change the weight given to the different functions in real time to better reach our goals. However, our methods are not based on these different kind of predictions that folks are working on for visualization. We are not using that yet.

Richard C. Alkire, University of Illinois, Urbana-Champaign: In addition to the chemical engineering department, I have an appointment at the computing group at Illinois, and I play the piano. I have been thinking of your work and about the way the eyes see data visually when playing the piano. The fingers touch the keys and the muscles drive them. There is integration to a considerable extent, but it is all connected to the brain.

We have a very large and still growing computational infrastructure in the United States, and we have fingers and eyes and data coming together to solve problems or create solutions. Could you comment in a forward-looking way on how all of these pieces will actually be integrated, how the data will be structured, so the most people can access the data in a proper way on computers for which they weren't originally intended and compiled, and how it can be accessible in a way that allows us to solve problems over and over and learn from them, just as you have envisioned? What is needed between all those fingers and the nerve endings that you have described and the brain that coordinates all the pieces that keeps them straight?

Venkat Venkatasubramanian: Certainly we are nowhere near such a level of complexity. That would involve database management and security issues, which we are not looking at right now. Eventually, when these kinds of systems are sitting in companies and institutions, both issues will be somewhat important. We have a long way to go to reach that point.

10

The Tacit Economics of Modeling: Indifference Curves that Should Defy Indifference

Michael Schrage[1]
Massachusetts Institute of Technology

The United States has had an explosion of innovation opportunities. This "opportunity glut" creates a need to explore both the meaning of innovation and the role it plays in letting firms profitably differentiate themselves in the marketplace.

As someone trained in economics, I have always been struck by organizations that had perfectly good, rational tools for getting the job done but that continued to act irrationally, counterproductively, and seemingly inexplicably. What legitimate reasons could economists develop to explain that organizations often ignore good technical solutions to their problems?

Assuming that an innovation opportunity can be modeled in some useful and meaningful way, I am interested in exploring how people behave with respect to the economics of and the tradeoffs associated with modeling. My particular emphasis is on innovation behavior, looking not at how people think but at how people behave. Actions speak louder than words. Actual behavior is more eloquent and revealing than rigorous analysis.

Explicitly, I look at how people interact around iterations of representations—or, in plain English, how people behave around versions of models. That behavior is the essence of innovation. This begs a simple proposition: when we transform the economics of modeling, prototyping, and simulation, we inherently transform the economics of innovation. Make modeling faster, cheaper, and easier and we surely change the economics of innovation. We incent different iterative and innovative behaviors.

To appreciate the real impact of this transformation, we need to annihilate some of the myths that the information technology domain has inflicted on us. We need to avoid remaining victims of a "data-driven" vocabulary.

The first myth I want to expose is what I call the Big Lie of the Information Age, which is that as we change the quantity and quality of people's information, we change the quantity and quality of people's

---

[1] Michael Schrage is co-director of the MIT Media Lab's eMarkets Initiative and a senior advisor to the MIT Security Studies Program. His research focuses on the role of models, prototypes, and simulations as essential media for managing innovation and risk. His book, *Serious Play* (Harvard Business School Press, 2000), explores the economics and ethology of modeling within organizations.

behavior. As reality attests, that is demonstrably not true. In fact, if we change the information in organizations, we oftentimes do not change people's behavior, because people frequently ignore or dismiss the information given them.

Consider an informal *gedanken* experiment: I asked people to choose between a tool that offers an order of magnitude—10×!—improvement in managing all the information that goes across the desktop, phone, personal data assistant, cell phone, and Web, or a tool that offers a 20 percent—0.2×—improvement in the ability to persuade one's bosses, colleagues, and subordinates. The overwhelming majority of people consistently chose the persuasive tool. So what is the real issue most people face in their organizations? Information or persuasion? The role and rhetoric of models and simulations for persuasion in organizations have very different design emphases and sensibilities compared to the models for informing organizations.

So the common reaction is to question how to develop models, prototypes, and simulations that make a person or his group more persuasive within the organization. That is a legitimate design question. Using myself as a beta site of one, I realize I don't like being persuaded by others. However, I consider myself open minded enough that I am happy to persuade myself. The design challenge? What is a better investment? A tool that helps a person become more persuasive or a tool that helps that person's customer, internal or external, persuade themselves.

I submit that building models that enable people to persuade themselves is a different design sensibility challenge from the designing of a model, prototype, or simulation that makes a person or group more persuasive to others. Those kinds of design parameters become more important rather than less important when we talk about accelerating the pace of innovation within an organization, industry, or market segment.

Why are these kinds of design questions more important now than even 5 years ago? We are rich. We have computational wealth and power that transform both the economics and rhetoric of modeling. However, I fear that we are using the wrong unit of analysis for the assessment and measurement of computation-driven influence. People talk about Information Technology and the Information Age, but I think that modeling, simulation, and prototyping of innovation are, frankly, not an information management problem.

Instead of "Bits" management being the issue here, the real source of wealth is "Its"—Iterations. Innovation wealth is a function of a shift from Bits to Its. Instead of managing information, how do we better manage iteration? The new wealth is our ability to iterate and perform more iterations per unit time. What do we do with that as individuals? What do we do with that as teams? How do we use the opportunity to manage more iterations per unit time as a vehicle to reduce coordination costs and transactions costs? How do we create these technologies as a vehicle to facilitate communication of innovation and management of innovation within organizations? The answers to these questions determine how well—and how poorly—organizations will iterate to innovate.

Please make a conceptual leap with me. We know what financial capital is; we understand and appreciate human capital. We hear more and more about social capital. I would like you to think of the explosion of computation-driven iteration to be a form of Capital. Iterative Capital.

So we need to ask ourselves what is our ROI—not Return on Investment but Return on Iteration. What do we want to accomplish as innovators? What kinds of attributes are we iterating around? What is it that we are really trying to learn as we iterate to innovate? Are we interested in the development of a particular structure or material or that structure or material in a certain kind of a context? How do we manage the return on iterations? Ultimately, the more choices you have, the more your values matter.

The Capital Asset Pricing Model and other financial theories around diversification offer useful insights. If I give you a million dollars, you are a fool to invest in venture capital. But if you have a billion dollars, you are a fool if you *don't* invest in venture capital.

As we transform the costs of iterations, we have to transform our investment profile. Being wealthy means you have more choices. If you have more choices, your values matter more. In the end you have a portfolio of iterations. How do we want to manage those iterations?

All the previous information makes sense on a "rational" analysis basis, but as we know, for every buyer in the stock market, there is a seller. These individuals and institutions have different risk profiles—they have different expectations of the future.

We have to look at real-world positive behavior as opposed to normative behavior. For organizations the true test of a model is not how well it works, but how well it is used. There are a number of very good reasons why organizations behave very irrationally with respect to their investment in models. There are three examples of this poor behavior: the golden goose, the magic mirror, and the stone soup.

Chemists are inevitably asked by organizations for better models for problem solving. In theory all employees in an organization understand the problem and are therefore capable of building better models that produce better answers. In reality everyone does not understand the problem and its underlying issues, so a model can be improved in two ways: by changing it to produce a better answer, and by changing it so that the model is more accessible to all employees. The modeler chooses from which direction the greater return for the effort comes. Ph.D. chemists are very good at communicating with other chemists; however, communication with business people is difficult. This poor communication usually results in increasingly better models that become increasingly less accessible to nonscientists.

If I offer people the choice of either golden goose eggs or the goose that lays the golden eggs, most choose the goose. Unfortunately, that is the wrong choice because there is no information on how much the goose costs, whether it's a mean goose, or whether it costs more to take care of the goose than the value of the eggs.

There are many organizations with many managers, particularly on the business side, who don't want models, prototypes, or simulations (the goose)—they want the answers to their questions (the golden egg). The models that have been built in the past are generally not engines of innovation. They have often been treated as a necessary evil and are merely overhead for finding the answers to the organization's questions. Which organizations are just building models that are technically interesting but don't generate answers about the economics of production for an innovative material, and which are investing in models to get answers?

The issue is similar to the goose and the golden eggs because it is usually unclear in an organization what the investment in a model is actually for and whose needs it meets. For example, is the investment in the model to get answers for only the next 6 months, or is the investment to build a more robust model able to meet changing needs and economics?

There is no reconciliation or discussion of that because the answer-driven people don't care about the model; they only care about the answers—they want the golden eggs. The chemists or modelers are only interested in the technical elegance of better models and do not understand the business aspects—they want the goose. There is no honest discussion of the distorted economics because one side of the house thinks they are investing in answers while the other side of the house thinks they are investing in a medium. So are you a golden goose organization or are you an eggs organization?

Then there is the magic mirror. You can stand in front of the magic mirror and ask it to make predictions: What am I going to look like 5 years from now if my lifestyle remains the same? What am

I going to look like 5 years from now if I exercise every day and eat right? What am I going to look like if I let myself go? What do I look like in a hat? What do I look like in used clothes? What will I look like if I get liposuction?

The mirror is smart enough to understand the questions. Ordinarily the interfaces of modeling and innovation are not so well defined. How much time would you spend in front of that mirror? What questions would you ask the mirror? What questions would you always ask the mirror? What questions would you never ask the mirror? Which images would you archive and preserve? Which images would you make sure once you saw them you never generated again?

But that is the gutless question set. The gutsy question is as follows: Do you take your significant other in to see your images on the mirror with you? Do you give your significant other the right to ask questions of the mirror about your future appearance, based on suggested changes? Some people would never want their significant other to see the worst-case scenario because it might frighten them away.

Conversely, some people would never want their significant other to see the best-case scenario because that is an unachievable ideal that they do not want to live up to. It requires no huge conceptual leap to realize that this is exactly the situation we are approaching in computational chemistry. Do we want our key suppliers to look in the mirror? Do we want our customers to look? What questions do we want our customers to ask or not ask? What are we prepared to do collaboratively and what not? The technologies for doing all of these things exist. The purpose of the information highway is not just to drive to the solution of a problem, but it is to drive to a location to share with other people, to ship goods to other people. There is a related social context, which leads to the stone soup story.

A vagabond puts a stone into a pot in the middle of a town. When asked what he is making, he replies, "Stone soup." The townspeople wonder how he could be making stone soup. "It is wonderful soup, but it can use some carrots every now and then," the vagabond says. So people volunteer to bring the carrots in. The stone soup is a way to get other people to participate and invest in the meal. One of the major problems with modeling infrastructures today is that they do not invite others to participate, and they do not run simulations that are capable of embracing and integrating other people's data. An invitation for them to cook and collaborate with you is needed.

What is so ironic is that collaboration can occur even with bad models. Often times people with very good models have more trouble collaborating than people with lower-quality models because the former group feels that their model is good without additional perspectives or input—it is presented as a *fait accompli*. The modelers have the answer, and they are trying to persuade you that their answer is right. The people with the mediocre models have a greater incentive to collaborate and pay more attention to a customer, a supplier, or other participant. In any collaborative context, business or academic, a less efficient model often invites greater participation by others than a superior model.

At the MIT Media Lab, the really skilled graduate students and professors stand out because they don't "show and tell." They "show and ask." They use a demonstration, model, or prototype to elicit the ideas and insights of others as opposed to convincing others of their perspective.

The people who really manage to build funding, create community, and obtain the interest of venture capitalists all use participatory styles. With this in mind, what are some actions to take that will manage some of the pathological issues associated with models, the things that you could actually use in an organization to leverage the modeling simulation, and the prototyping infrastructures that you have to manage for the innovation process?

First, is the model or simulation designed in a way that it can be used by a key customer or key supplier? Are you designing for accessibility? The use of your model by others is an excellent test of the model's accessibility. When one of your key customers or key suppliers is using your model in a way that you never expected it to be used, your model has succeeded. It is tremendous to learn that your research and development has become market research.

A model can be thought of not just as an interaction engine or an information engine but as a marketplace, where people trade ideas. For example, one group in the organization can only have 30 iterations, another group can have 60 iterations, and they swap and trade between them.

Second, a good model, prototype, or algorithm should attract the attention of other people. The modeler does not need to be charismatic; the work should be charismatic. This is one of the advantages of very good models. The sign of a superstar is not scoring a lot of points but making everyone else on the team play better. In an organization, models and simulations make people think better. They may not even be particularly good models, but they may force a more creative, intriguing, provocative, and useful kind of thinking. Look at how people behave around models, rather than just how models themselves behave.

Finally, the "80/20" rule is absolutely true for models as it is for most things in economics: 20 percent of the model generates 80 percent of the usage. For any model it would be fascinating to know what functionality of the model, what 20 percent, generated 80 percent of the usage. Look at any piece of computer software—only a tiny fraction of the functionality gets used. As you evolve a model, what is the 20/80 ratio? Such auditing is not just targeted experimentation but targeted evolution of how the model is actually used relative to the declared potential of the model.

This kind of introspection will give you tremendous insight into the economics and culture of modeling within an organization, because the real problem is how the models are actually used rather than how they are actually built. There is a utility problem with models more than a design problem. At the core of that is the issue of the economics—distorted economics that lead to distorted behavior. As we become smarter about the internal economics, we will become smarter about the usage of our models and our tools.

DISCUSSION

Joseph S. Francisco, Purdue University: There is an area of economics that is very exciting called experimental economics. Some of the people here from industry are thinking about taking modeling results and utilizing your suggestion of trades by playing it out within the experimental economics scenario of looking at the outcomes. This could help to guide their investment decisions after seeing the outcome of their initial choice. Is that what you are trying to suggest?

Michael Schrage: That is one of the things. In fact, I am shocked and impressed that you raise that question because there are two lines of thought. There are experimental economics and behavioral economics and finance,[2] which have begun to converge. There is work currently being done on the modeling of experimental economics in finance and investment. It would be very interesting to pick those same economics principles and apply them to the planning of a chemical or chemical engineering experiment, treating that as the marketplace. Such research could discover the tradeoff between rational investment decisions and distortions in behavior because of prospect, theory, biases, cognitive biases, and the like. The convergence of experimental economics and behavioral finance, I think, is going to play an enormous role in ultimately shaping how research and development institutions both make and justify their investments.

Kenneth A. Pickar, California Institute of Technology: As a manager, I find what you say to be one of the great attributes of terrific ideas: it is blindingly simple and has the virtue of matching my own

---

[2]Shortly after this workshop, the 2002 Nobel Prize in economics was awarded to Daniel Kahneman and Vernon L. Smith for research in these two areas.

prejudices. I will give you two examples. First, the standardization of tools is extremely difficult. This does not apply only to models but also to computer tools for mechanical engineers or a piece of software. The standardization of tools is in everyone's best interest once it is done, but to convince someone that the tools they are using are not as good as the one you want them to use is nearly impossible and requires almost an autocratic approach. Sometimes we will find ways of making it not work.

The second great example is the election in Florida. Here you have intelligent people given the same amount of information and, depending on their predilections, they were convinced that they were being robbed by the other side in a dishonest underhanded way. They were equally passionate about their side of the conflict, and yet in other ways they seemed like reasonable people.

This fits into the context of accepting new information. The idea of making tools very useful, as distinguished from very cool, is important. The part with which I had some trouble is the optimization around bits because my rudimentary knowledge of economics says that you look for scarce resources and try to make them go the farthest. You could iterate a thousand times.

Michael Schrage: No, whenever you have an abundance of a resource, lots of time, lots of people, there is an issue of waste, and when we are dealing in a competitive environment, you need to know when you have hit the point of diminishing returns.

We have got to be careful—you can go past the point of diminishing returns by making sure that every contingency is planned for with Monte Carlo everything. We must be aware that there is such a thing as diminishing returns and that, when a resource is growing, we want to make sure that we don't go down a groove or a rut and that we are at least aware of the tradeoffs.

Kenneth A. Pickar: I will give you a perfect anecdote. Part of my job is to see how engineers are actually using their tools. There was one engineer who was using a particular mechanical engineering tool and had just completed 150 iterations. It turned out that the results had hardly changed between the 20th iteration and the 150th iteration. So I asked him why he had continued to perform iterations. He answered that the time it takes to finish 150 iterations was the amount of time we had allocated in the program to do this.

David E. Nikles, University of Alabama: I have a management problem. I run a team of faculty, which is like herding cats. While I am a grubby experimentalist, we have modelers who talk about the elegant math they will use to do this and that and model real-world phenomena. The models are never finished in time to be useful for the experimental project. I see a fundamental disconnect because I cannot use the model myself for an application. On the other hand, maybe the modelers should be doing some experimenting. How do you manage that? I think the experimentalists have to meet them halfway somehow.

Michael Schrage: Part of the problem goes back to Ph.D. chemists' lack of ability to communicate with people in their subdiscipline. This is problematic, particularly since there is a trend toward funding collaborative and multidisciplinary projects.

If modelers cannot design an interface for the model that other people in the group can use, they fail because they have not produced a model; they have made a black box. Opacity should not be acceptable for any models designed for an interdisciplinary or multidisciplinary setting.

David E. Nikles: I think there are many things we can learn from modeling. However, the modelers always model what I already know and they never tell me something that I didn't know.

Michael Schrage: There are two important characteristics of models: accessibility and generation of novel, sometimes counterintuitive, results.

Mary L. Mandich, Lucent Technologies: I know the employee at Bell Laboratories who wrote Troff, which was an early but very powerful word-processing program. It was so complicated that every secretary who came onboard had to spend weeks in training learning it. There was a computer consultant in every building who you went to when you had trouble. One time I asked this man why they did not take some time to make the program easy to use, and he said that that would have been harder than building the original program. It seems to me that you are telling us not to spend time abandoning the complexity but to spend time on the equally hard stuff.

Michael Schrage: Right. It may be not be appropriate for certain kinds of model builders to focus on user interfaces because that is not how modelers are trained. There are two things going on. There is designing for the better model versus better access. An experimentalist might get useful information from what the modeler would consider to be junk.

There needs to be that kind of negotiation between modeler and experimentalist. Additionally, there are interesting real-world situations that will make the modelers think twice about what the meaning of elegance in the model actually is.

It is necessary to use the model as a medium to manage interaction between the positive and the normative folks. That is one of the underplayed aspects of the problem because the modelers are optimizing it for their community rather than optimizing the models for interaction. You can do this in academe. In business it fails completely. The real question is how do we want this community of collaboration to evolve, and we can adjust our investment in the modeling infrastructures based on that answer.

Richard C. Alkire, University of Illinois at Urbana: Your economic examples here are good ones because they take us out of our box and make us think from another perspective. The impact of all of this information, not only on science and engineering but on our ability to determine the way we will live in the future, is so great that, like economics, it goes beyond economic principles. It gets into issues of ethics.

People understand this in the economics where economic issues become transformed into questions about the capabilities of economic instruments to do good things, especially how they pass from one culture to another and empower people to assemble the resources they value. Architects also think about these things when they create structures that shape the way people live their lives day to day, moment to moment. There are architectural features that have a sense of comfort, whether it is an interesting entrance to a house or a garden bench next to a wall. We all know when we are happy and when we are not happy, not because of economic decisions or architectural designs but because we feel comfortable. Turning now to the Information Age, could you comment on the process by which we might learn to design our living space so that we can feel comfortable in the presence of so much information?

Michael Schrage: That is an extraordinarily difficult question to answer; therefore, I am going to oversimplify it to a level where I am comfortable answering it. There is a major ideological battle going on between the normativists and the positivists. That battle is fundamentally based on the core of economics—how people should behave—which exemplifies rational choice versus actual behavior. Theories like experimental economics and behavioral finance represent some effort to arbitrate normative expectations with positive observations.

I have made a slow and eventually accelerating migration to the positive side because people don't behave rationally, how they are expected to. There are two superb books addressing this. *Image of the City* by Kevin Lynch discusses the mental maps that people have. In this book, cognitive maps and spatial representations are tied into the architecture of city planning, which is directly relevant to many issues in design. The other book is *How Buildings Learn: What Happens After They're Built* by Stewart Brand. *Serious Play: How the World's Best Companies Simulate to Innovate*, authored by me and Tom Peters, looks at the issues associated with models, prototypes, and simulations and the culture of models, prototypes, and simulations in organizations.

Richard C. Alkire: I would like to mention that my own remarks are based on a book called *A Pattern Language: Towns, Buildings, Construction* by Christopher Alexander.

Michael Schrage: His work is superb. Incidentally, the ideas in *A Pattern Language* have been reappropriated by the object-oriented software designers (architectural design rules, laws of form, and the like)—truly technology transfer.

David J. Soderberg, BP Chemicals: I believe that the psychology of model application, technology management, and the business interface are more important than economics. One of the challenges technology managers have is interfacing with businesses. For example, in process modeling a group of modelers will talk to a group of process engineers about a model, but because of their common backgrounds, they drive each other for a better model, rather than one that is meaningful to the business.

Michael Schrage: The model is often designed to coordinate how the company behaves as opposed to how it interacts with clients or customers. Your example reinforces the fact that there are different design emphases.

Unless they are designing the model as a medium for communication, in addition to a medium to improve their problem-solving capabilities, they have failed in their professional responsibilities.

Walter G. Copan, Lubrizol Corporation: I wanted to thank you for your insight on the subject. Your comments on these issues have resonated with many of us. We have certainly begun to see the power of the potential represented by the interactions between human beings and models. This impact can be seen in new kinds of customer-supplier relationships as well as partner interactions. Because shared models create a deep, rich dialogue that provides profound insights, a much higher level of understanding is possible.

I would like to ask you a couple of questions about the power that models have to increase interaction and understanding. First, what are the elements of success you believe are critical to achieving the ultimate benefits of interplay between organizations and the full use of the power of their models? Also, what do you believe will happen to the future of customer-supplier or partner interactions as a result of having models available that are commonly built?

Michael Schrage: Well, those are both great questions. For models to make the maximum impact possible, we need the different incentive structures to encourage model usage. There is a very simple rule: using a model should be easier than not using a model. The real question, then, is if I learn to use this model, how long will it take me to feel like I am getting a real benefit from it? That deals directly with user interface.

Employees need an incentive, a reward for using the model. I believe that most large organizations should set up two kinds of prizes, one for the group or team whose model gets used the most by other people in the organization, the other for the person or group that does the best job of stealing somebody else's model.

Walter G. Copan: The second question is "Where could modeling lead us ultimately in terms of customer-supplier and -partner interactions?"

Michael Schrage: To answer the second question, I believe that business intercourse and design intercourse will increasingly be mediated by models. A wealth of doctoral theses will be done along the dimension of in which industries do the vendors use the customers' models, and we are seeing the supply chain management, and in what industries do the customers use the vendor models, which we oftentimes see in aerospace or high-tech industries, where the suppliers' competitive advantages are disproportionately more sophisticated than their customers'.

There are information asymmetries in certain industries regarding whether the supplier or the customer has a competitive advantage by using models. These information asymmetries will be reflected in the shared spaces, and models will become the media for collaboration. They will become the bridges and the glue between disparate organizations.

11

Successful Innovation Starting in an Academic Environment

Richard K. Koehn[1]
Salus Therapeutics, Inc.

My career for the past 25 years has been managing innovation of technology in universities and bringing it to commercial development, either by forming new companies or by making licensing deals. Managing innovation means managing the infrastructure that accelerates both the rate at which innovation or discovery may occur and the rate at which it may successfully flow into the private sector. I will focus on policies and infrastructure that can accelerate university discovery and innovation. Although I will not focus on the chemical sciences, everything I say is extremely relevant to the chemical sciences and to any science and engineering discipline.

I am concerned with the factors that enhance the rate of discovery and the yield that we garner from the discovery process. We tend to think of this process in very simple terms: there is discovery, something happens, and it has economic impact. In fact, a lot of institutions are or have been of the opinion that if you pour more money in at the front end of the process, more comes out at the back end. No matter how inefficient the process may be, we know that is simply not the case, and that there is, in fact, a large series of events and activities that occur in the course of the transition of an invention, including technology transfer and corporate development to some form of economic impact.

We tend to think about discovery and innovation in linear terms, but the process is completely nonlinear. There are all kinds of nonlinearities—for example, the technology transfer process of the feedback that is built into the culture of an institution, who gets rewarded for what behavior, and whether there is economic benefit to the individual participants in that process. The nonlinearity of innovation makes it interesting and difficult to manage while trying to enhance the efficiency of the process.

The intermediate phase of the innovation process, where innovation is transformed into a product with an economic impact, is the stage at which action by an institution can enhance the economic impact of the discovery. Three factors significantly increase rates of innovation during this stage. The first is the

---

[1] Richard Koehn is currently president and CEO of Salus Therapeutics, Inc., an emerging biotechnology company. Prior to holding that position, he was vice president for research at the University of Utah, professor of ecology and evolution and dean of biological sciences at the State University of New York at Stony Brook, and director of the Center for Advanced Biomedical Biotechnology for New York state.

financial factor—how much investment is made in the discovery process? The second factor is administrative policies and practices of the university. These policies are usually a potential hindrance to the innovation process. The third factor that affects the rate of innovation is cultural. This of course is not entirely separate from the administrative factor, and it describes how the faculty (individually and collectively) and the corporate institutions see themselves as members of a larger community.

There are four important sources of research support: private, governmental, intramural, and corporate. These funding sources decrease in flexibility of use in that order, with corporate funding being the least flexible.

Intramural funding is monies mobilized, identified, and deployed by an institution for a specific purpose. It is the most significant type of funding within the institution's control. Therefore, I will spend most of the talk on this intramural funding. What is it about intramural research programs that makes them important and successful? They increase an institution's ability to leverage itself into more competitive positions, to more successfully garner federal research support, and to more successfully interact with corporate entities outside the university. Therefore, the way the institution mobilizes its own flexible internal resources to support faculty research is a decision made very carefully.

To be successful, an intramural research program should be innovative. It should be feasible. It should be something that can be accomplished within the realm of possibility of the time and resources defined in the program. It should be timely and affordable and should create market demand. Universities are often very bad at identifying how well a particular proposed idea fits within the larger context of the private sector and whether it has any chance of being competitive, productive, or desirable in that arena. An intramural research program ought to ultimately generate some revenue for the institution, at least from increased research funding.

In effect, these programs are local venture capital funds. The funds are invested in projects, which, if successful, will result in a technology that is either licensable to a specific market or that will provide a platform technology for a new start-up company. The investment of these funds is made by the same criteria that any private investor would make in a program. How much is it going to cost? How soon am I going to get a return on my investment and what sort of return on investment might I get? I would like to look at two of these programs.

The first program is the Innovative Technology Development Program at the Center for Biotechnology at Stonybrook University, started in 1983, but unfortunately there has not been a thorough analysis of the program's productivity. When I called Stonybrook to get some data on the program, I touched a raw nerve. There seems to be a reluctance, which is not unique to Stonybrook University, to look at these programs with a hard eye and to determine whether innovation programs are achieving the goals that institutions originally set forth. The programs are ongoing with available funds that are dispersed every year. Even so, it should be determined whether each program is competitive in the desired way. The Innovative Technology Development Program was intended to enhance corporate sponsorship for biotechnology research, and these funds were awarded to the institution out of the Science and Technology Office of New York state, which deals with economic development. These are outside funds that are specifically intended to enhance the interaction between the university and the corporate biotechnology community. Thus, the number and quality of interactions are the metric that should be used for measuring the success of this program.

The Innovative Technology Development Program is an internal grant program focused on biotechnology with the intent of supporting projects that, if successful, will result in technological innovation of biotechnology and a leveraging of those funds by partnering with a biotechnology or pharmaceutical corporation. To date, the program has distributed about $8.5 million; each project can receive between $40,000 and $70,000 per year for multiple years.

Of the 95 current license agreements that the university has, 66 of these (70 percent) were funded by this program. This is an impressive statistic. It has, therefore, generated 3.7 licenses per year and has produced about $2.5 million in royalty revenue. The program has cost $8.5 million so far, and it has generated $2.5 million in royalties. If the development of the pharmaceutical RealPro can be credited to this program, the royalty value is over $50 million. The lesson to be learned here is to structure an intramural funding program in a way that serves the fundamental goals of the program before the program succeeds, because once you do, everyone wants the money. If it is really a program to enhance innovation, it is important to generate revenue and invest that revenue back into innovation.

The investment of intramural funds for a particular focused purpose has, in fact, had multiple effects in leveraging the return on investment of those funds, both from the corporate sector and from royalties. The leverage factor for the Stonybrook program has been significant in terms of both corporate support (the original, fairly narrow intention of this program) and a larger research support that has been derived from federal sources. Intramural programs have the ability to leverage a significant return on that investment if appropriately managed.

The other program I'd like to speak about is the Technology Innovation Program that was started at the University of Utah in 1994. The investment so far has been $2.7 million and the annual awards for individual projects are $35,000 maximum for 2 years. There have been approximately a dozen projects per year funded since the beginning of the program. The productivity is all in the early years. That is, the only sales that are being made from this program are those from 1994. Thus, the average time between investment of a project and return on investment is 8 1/2 years, an important point for university administrators who seek a quick return on investment in intramural research.

It is important to remember that technology innovation programs are not simply a quick way to make money, which many university presidents would love to believe. For investments generally and these intramural programs specifically, there is a significant time lag between the investment that you made and the return on that investment. This is a long-term program.

The conclusions of these two case studies, which are similar to others I know about anecdotally, is that arguably all the cases represented here are above baseline innovation: the patents that have been filed, the companies that have started, and the licenses that have been negotiated are really metrics of productivity that *would not have happened in the absence of these intramural programs.*

It is hard to tell whether the return on investment from federal funds exists or not. It is nevertheless unequivocal that both the rate of discovery and the rate of innovation at Stonybrook University and the University of Utah have been substantially affected by these two intramural programs. Intramural programs have increased discovery and impact. Fifty percent of these projects have generated patents or patent applications. That is much higher than typical university research projects sponsored by federal grants. There is a significant time lag, as I mentioned. The return on investment from royalties did not pay costs in either program, and yet there was an overall significant leveraging of funds of different kinds.

I believe that the return on investment from royalties will pay the overall cost of this program eventually in both cases. The final conclusion about intramural programs is that more aggressive management accelerates innovation. There are ways in which institutions can manage these processes that make them more successful and productive.

When projects are managed more aggressively (more closely), they are more often monitored by individuals capable of measuring the progress on these projects. This makes the projects more likely to achieve the milestones that were set. Thus, the three elements that I would like to mention that influence the rate of progress are technology transfer management, management and structure, and the university culture.

Technology transfer management, which involves tracking intervention, produces more information on licensing leads. Closer tracking makes the licensing opportunities for that technology more obvious, and a closely tracked project is therefore more likely to meet the commercialization milestones and more likely to identify new factors to produce innovation. Typically, a faculty member sets out a line of investigation with specific milestones, but during the course of that investigation, discoveries are made that can be more important as an end point than the original end point. Project management is an important issue. For example, adequate budgeting to get patent disclosures and to process those patent claims is critical to the overall operation.

Some institutions use faculty committees to vet inventions and progress, which is very inefficient and unproductive. Decisions about projects should be made by professionals who invest funds and seek a return from those investments. One of the characteristic differences in the evaluation process of these projects is that unlike general grants the evaluating panels argue from the private-sector perspective and will simply terminate a project when milestones have not been made.

When a project is not successful and the return on investment is inadequate, the best decision by program managers is to terminate the project. Such a decision-making scheme, although generally uncharacteristic of university research, is critically important in the management of the funds invested in innovation development projects. These programs also need to be responsive to faculty needs and initiatives by encouraging partnerships between the investors and the scientists. Before the program begins, management must also decide between central and decentralized decision making and the subsidiary corporate structure for management of these projects.

University culture is one of the more critical elements in managing the overall technology innovation process and the productivity of investment funds. In the early 1980s, when I first put the Center for Biotechnology together at Stonybrook and began to allocate funds for these projects, the faculty said that they did not do applied work because it was not very interesting: "I am a basic research scientist. I want to do exciting new innovative work." Perhaps the response I am describing is more characteristic of the life sciences, rather than engineering or chemistry, but regardless, there has been a cultural evolution in the university since that time.

Initially, there was a real distinction made in the minds of biologists between applied, commercially oriented work and exciting innovative basic research. Two things have happened that have accelerated the cultural evolution of attitudes away from that point. The first is that research faculty has found that it is possible to do exciting innovative life science research and have elements of that work that are extraordinarily important in terms of a variety of practical applications and commercial exploitation. Second, it has been helpful that a few professorships have been created this way. That is a nice role model that some faculty have seen.

Consulting policies are not static and permanent, but are key and critical to the protection of ideas and the ultimate exploitation of those ideas. The policies that govern how faculty can interact with elements outside the university and share their ideas is important to the overall way in which technology is managed and captured within the institution.

Whether faculty inventions are licensed to faculty entrepreneurs varies from institution to institution. In some places, if a faculty member has invented a core technology and wants to start his or her own company based on that technology, he or she has to do that outside the university, not as a faculty member. At other places a faculty member can start and run the company within the university so long as the chair is not disturbed by it and it will spin out eventually. When the distinction between the role of faculty and the role of the CEO of a company becomes blurred, it leads to all kinds of extraordinarily difficult management problems. Conflict of interest policies are also a problem for management.

Equity-sharing policies must also be addressed. To what degree should faculty entrepreneurs be able to share in the equity of a start-up company? To what degree should each institution take equity in the start-up of its own technology? To what degree does that create a conflict of interest structure? Does having a patent hurt you? I have seen very successful faculty entrepreneurs devalued by the academic evaluation environment because of the idea that, if they are applying for patents, their work must not be scholarly. The extent to which we see success in technology innovation being a part and parcel of the overall institution is a cultural issue.

Is innovation simply a means of generating revenue for the institution, or is innovation seen as a reflection of a real change in the corporate university, which requires a change in policy and procedure and, ultimately, culture? The university president can make a difference in the promotion of innovation, but the tone of the administration, the institution's infrastructure, and its policies more heavily reflect the institution's ability to promote innovation technology development.

DISCUSSION

Nancy L. Parenteau, Organogenesis, Inc.: I would like to ask you to elaborate a little bit more on the conflict of interest.

Richard Koehn: Let me ask Francis to respond.

Francis A. Via, Fairfield Resources, Inc.: Universities have been changing the general structure, purpose, and activities of research with respect to intellectual property, as you have clearly outlined. Recent legal rulings on biotechnology, high-throughput screening, and combinatorial chemistry patent law have focused on issues associated with the research or experimental use exemption for intellectual property.[2]

The most recent interpretations indicate that if you experimentally practice technology that is patented with the intent of expanding scientific knowledge or conducting curiosity-driven or basic science, you are free to do so. However, if you are conducting that research with a profit motive in mind, if your intention is to develop a product for commerce or intellectual property for license, you do not qualify for the research exemption and are subject to any patent covering the technology in question.

I wonder how universities are beginning to address this issue because their research direction or intent has changed with the goal of gaining intellectual property coverage.

Richard Koehn: I do not think the universities have any idea how intellectual property laws relate to the general research mission of the institution or its desire to exploit the fruits of that research through commercialization. It is completely different when you are doing research in chemistry on a particular area and you see some particular applications in mind, but you are actually utilizing patented procedures or processes in that research. Have you violated the patent? The question of a patent violation in research laboratories is extremely sophisticated, and most technology transfer offices at universities do not know that the issue exists or how to think about it. Now that the universities are thinking about exploiting the commercial value of a project, they need to ask what process was used to produce the fruits of that project. That is a different level of sophistication.

---

[2]More information about the intellectual property exemption can be found in the National Research Council report *Intellectual Property Rights and Research Tools in Molecular Biology* (Washington, DC: National Academy Press, 1996). This is also available on the Internet at <http://www.nap.edu/readingroom/books/property/2.html#experimental>.

One issue being considered in Congress is that public universities currently cannot be litigated against for certain kinds of patent violations because they are public institutions. There was a proposal to change that.

One of the factors involved is that universities are evolving from an era when they had no interest or idea that their research was of commercial value to one in which they begin to make this realization. Some institutions became very sophisticated at the beginning because they recognized that these partnerships are a way to generate revenue.

Many companies are at least aware of this intellectual property issue. For example, in my company I have paid for legal opinions on the freedom to operate, because I am concerned about this issue, but I do not know of any university that has ever done that.

J. Stewart Witzeman, Eastman Chemical: It makes sense that you could define policies for what you called intramural work. That is seed money that the universities invested; the ownership is clear. Can you comment on corporate-sponsored work? There may be underpinning technology owned by the university and that is a real bone of contention.

Richard Koehn: It is a strong bone of contention and the only way to practically think about the problem is to explicitly specify in the contract between the company and the university how this work will be done and what results are expected. The ownership, of course, will flow from that.

I think it is very difficult to build a fence around a particular project within an institution and claim that all of the discovery that happened on the dollars provided by the industrial partner was within that fence. It is difficult to claim that there was no proprietary knowledge or knowledge owned by the university that impacted that discovery in some way because of the context within which the project occurred. It is not easily resolved, and the only way you can approach the problem is to try to delineate before the fact what it is you are going to do, what you expect to arise out of that, and what the relative corporate positions on ownership will be once that project is finished.

J. Stewart Witzeman: One of the things we talked about yesterday was that often the faculty and the industrial folks can reach agreement, but it is often university administration that is perhaps not as sophisticated.

Richard Koehn: That is because the faculty will often make an agreement with industry without considering anything beyond their own welfare. I used to spend most of my time trying to sort out such conflicts, but this has become less of an issue as faculty become more sophisticated and as institutions have become more knowledgeable.

The major problem that I had as the vice president for research is that I often entered into a legal corporate agreement between, for example, the University of Utah Corporation and Eastman Kodak Corporation, for the performance of certain tasks and certain activities. These tasks were wholly dependent on the activities of the university faculty, in whose interest I was signing this contract. If that contract were not fulfilled, I had no recourse as a corporate officer in trying to resolve that difficulty with my faculty.

I can send them to a faculty conflict of interest committee, where they will have a long discussion for 9 or 12 months and ultimately resolve that they do not understand these issues anyway. Truly, I think that the most significant management challenge at research universities today is this growing disparity between the emerging corporate university and the traditional university.

Its employees are essentially free agents with tenure over which the university corporation has no real control. I am not suggesting that somehow there is something wrong with the faculty. I am simply saying there is something wrong with the structure.

The question is what is the institution able to do if faculty do not fulfill their corporate contracts. There is no recourse. The universities are simply not inclined to fire tenured faculty for violations of that kind. In some cases the institution reimburses the corporation. In other cases the institution negotiates some exit strategy that both parties can live with. In some cases real legal recourse is taken. This is a major problem.

Francis Via: I want to indicate, as Rick Gross of Dow pointed out yesterday, that many of these industry-university partnerships are based on choosing the right people and on trust. We have had more than a hundred programs with universities and have never encountered such problems. One item that demonstrates this trust was the agreement we had with the California Institute of Technology (Caltech). There professors do not sign the contract, but they are intimately involved in the negotiation and discussions. At one point, we wanted a statement that Caltech would not publish the results unless we had 6 months to review them and to secure the patents.

Caltech would not agree to that because it valued the intellectual freedom of the professors. Instead, the professors agreed to send us manuscripts and to wait for our signal to publish over a handshake. It worked perfectly. It is a question of trust as well as association with all the legal ramifications.

Richard Koehn: I couldn't agree more that it is a question of trust. The problem is always when that trust breaks down. The real problem is the lack of recourse.

12

Panel Discussion

Andrew Kaldor, ExxonMobil: This last session is an experiment to turn the meeting over to the audience and let you provide the appropriate ending. We wanted to see whether there was an interest in some impromptu presentations stimulated by the discussion, and Parry Norling has volunteered to give us a brief discussion on innovation in the public sector. When Parry is finished, the speakers who are here can join us at the front of the room. I have some questions that have been given to me by the audience through notes and discussion. I suggest that we walk our way through those questions. I will moderate the discussion, and I remind the speakers to address these questions as their interest dictates. Also, I encourage the speakers to ask questions of each other, since I am sure there were some issues of interest raised.

Parry M. Norling, RAND: What can we say about innovation in the public sector? A government agency recently asked RAND how it might be more innovative in all the work it does, not only in research and development (R&D), and requested some case studies of a number of public- and private-sector organizations known to be quite innovative.

We are trying to answer the following questions: Can a government agency be more innovative? Are the best practices and lessons such as we have been discussing today useful? Are lessons learned from innovative organizations in the private sector transferable to the public sector and vice versa?

The barriers to innovation in government have been recognized for centuries. Many years ago, Machiavelli wrote: "There is nothing more difficult . . . , more perilous to conduct, or more uncertain in its success, than to take the lead in the introduction of a new order of things. Because the innovator has for enemies all those who have done well under the old conditions, and luke warm defenders in those who may do well under the new. This coolness arises partly from fear of opponents who have the laws on their side and partly from the incredulity of men, who do not readily believe in new things until they have had a long experience of them."[1]

---

[1]Machiavelli. 1515. *The Prince*. Chapter VI.

Today's perils have been pointed out by Paul C. Light in his study of 26 innovative organizations in Minnesota: "Imagine the worst possible circumstances for sustaining innovation in the public sector. The external environment would have unrelenting turbulence and unending shocks. The non-profit and government organizations would be constantly guessing about the next crisis, thereby increasing the risks associated with investing whatever scarce resources they might have in innovation. There would be public cynicism. Collaboration among organizations would be discouraged by pitting one against another through categorical funding programs and by reducing the discretionary dollars for true experimentation. The worst possible scenario for innovation in the public sector would also have implementable internal structures and countless other barriers to innovation."[2]

At RAND we have been trying to describe and analyze systems and models for innovation in the select group of public and private organizations. We chose to study the U.S. Customs Service, the Food and Drug Administration, the Veterans Health Administration, Procter & Gamble, DuPont, and Marriott (and also had discussions with the National Institute of Standards and Technology and Rohm and Haas). We are still in the middle of this study and analysis and have given one briefing to the sponsoring agency with our findings. A number of interventions similar to those in the six organizations appear to be practical and may be adopted by the agency. Our report will be published in November and should be available on the RAND website.[3]

Andrew Kaldor: Will the panel members please join us up front? I will read the question from here. One questions was: Are there ways to overcome the distance effect on collaboration?

Elsa Reichmanis, Lucent Technologies: I have had some direct experience with long-distance collaborations. In one collaboration we had a joint R&D program with a company located in Switzerland, and in another we collaborated with a company located on the East Coast, only a few hours away from our location. In both cases the initial few meetings were face to face, which helped with the social aspects of the project. Because we had the chance to get to know each other, it was much easier to pick up the phone and talk to the relevant researchers in either organization to get information and understand what was going on. Daily e-mail communication helped us stay informed of progress.

Michael Schrage, Massachusetts Institute of Technology: One of the seductive dangers of technology is to believe that you can substitute bandwidth for presence. I think it is demonstrably clear that this is not true, and I find it ironic that we waste so much time on teleconferencing and video conferencing.

I think we need to define the word "collaboration." It is not a catchall. Do we mean an informal chat? Do we mean three people brainstorming around a particular model or representation? I am especially looking forward to seeing how different groups and different organizations develop a more refined and sophisticated definition of the word "collaboration."

Francis A. Via, Fairfield Resources, Inc.: The last few years at GE we focused on enhancing and improving long-distance collaboration involving global R&D teams working on specific technical goals. A system was being established to electronically monitor experiments that were being conducted at remote sites—for example, test trials in Bangalore, India, could be "observed" from laboratories conducting related work in Schenectady, New York. One concern rose rather quickly, that of the

---

[2]Paul C. Light. 1998. *Sustaining Innovation: Creating Nonprofit and Government Organizations that Innovate Naturally.* San Francisco: Jossey Bass.

[3]<http://www.rand.org/>

scientists needing "space" to review experimental results before a wideband broadcast occurs; others were personal concerns over the potential for displaying test failures on a daily and hourly basis. As we are witnessing, technology is increasing our ability to exchange information and to enhance collaboration. Nonetheless, there remain personal issues related to the methods for reviewing or controlling information.

There are many challenges associated with long-distance collaboration. As Allen Clamen pointed out, there is a growing variety of techniques becoming available, but few things are as useful as face-to-face contact on a periodic basis to build trust over time. Another important issue impacting collaboration is the economic environment: during an expanding economic environment collaborations are more readily fostered, while despite policies, strategies, or technologies, a contracting economic environment represents barriers and challenges to collaboration.

Venkat Venkatasubramanian, Purdue University: I want to follow up on Elsa's comments. All three projects I have worked on in the design area involved very strong collaborative work with our industrial counterparts with periodic exchanges of people. One of the advantages of being at a university is that the students can act as photons, mediating exchange. So we would send our photons out to spend 2 weeks to 2 months learning the details of technology transfer. Having the students as a medium to go back and forth made a big difference in learning the complexities of the problem and aided in transferring knowledge from our side into industrial practice on the shop floor.

Allen Clamen, ExxonMobil (retired): I agree with Michael that collaboration has many different forms. I am intrigued by the way needs and solutions find each other. I think the challenge we have at the very early stages in the idea management process is to find information within or even outside the organization that could make immature ideas more robust and attractive, which will persuade others that the ideas are worth pursuing.

I have heard that companies have done this by having a profile for each and every organizational member with their interests and their experience base on the company's Intranet. An idea that enters the database might be read by an individual and stimulate him or her to add to or enrich the idea and make it much more likely to have a successful evaluation and more significance.

Participant: I could give both good examples and bad examples. The best example I recall was when GE had a telecom product that was being installed in Japan at the same time we had a collaboration with Ericsson in Stockholm. A problem that would appear at the end of the day in Tokyo was sent to Lynchburg, Virginia, to be solved. The solution was then validated in Stockholm, and when the engineer returned in the morning in Tokyo, the answer was on his desk. It was very exciting. And no one told them to do that. The engineers just devised this technique themselves. Most of the great innovations don't come from great professors thinking about great thoughts, but rather from those for whom practicality is important.

One example of poor collaboration is the global failure of telecom to improve teleconferencing technology. Let me ask Michael about research at the Massachusetts Institute of Technology (MIT) Media Lab. Could you postulate a time when technology really will make a difference in long-distance collaboration? For example, what would happen if I had a screen right beside my desk that would show my buddy all the time, just as I would see a colleague in another cubicle. Of course, there is always the time zone problem. Regardless, it is currently unthinkable that at some point you should be able to have the kind of relationship with someone in Chicago, down the street, or more than 90 feet away that you would have with someone who is next door.

Michael Schrage: The transmission of presence research was funded by the Advanced Research Projects Agency (ARPA) at the Media Lab more than 20 years ago. We do have ubiquitous technology that allows a person to be with you all the time. They are called cell phones. There is no longer such a thing as an uninterrupted conversation.

I have found that even though we love face-to-face meetings, the most important thing in terms of productivity and collaboration is the transmission of the work as opposed to the transmission of the people.

It is very rare that you need to see the individual. However, you do need pointing capabilities, highlighting capabilities, and annotation capabilities. The notion of collaboration around work versus collaboration between individuals strikes me as one of the ways that we will see different tool sets evolve to support those kinds of interactions. Thus, while I accept your premise, I really believe that enterprises are going to focus on things like instrumenting experiments and design representations, as opposed to substituting for this kind of face-to-face interaction.

Participant: I am astonished that you believe the person can be separated from the work. I have many counterexamples. When I was at GE working on magnetic resonance imaging, we had Japanese collaborators who were part of a subsidiary of GE. They were not strangers, but they came to the United States, worked in our laboratory, and were seen at night making phone calls to Tokyo. All the wrong conclusions were drawn. We thought that somehow they were stealing our stuff and the situation was horrible. It was just paranoia on our part. You can imagine the lack of sharing of data that then ensued because people just didn't trust each other.

Elsa Reichmanis: I think that raises a point of needing to have trust and respect, in addition to communication among the group of people who are working together on a body of work.

Andrew Kaldor: I agree. I think face-to-face contact is needed to build up a trust. Then you can try to share information.

Lawrence H. Dubois, SRI International: There are three broad recommendations that I came up with to stimulate some discussion. One falls into the class of focusing on important problems and not just interesting problems. The focus should be on something that is big, important, and will make a difference. In the university setting, working on an important problem makes the research much more relevant to students because they are working on something that has some value. You have much more relevance to the government funding agencies, potential industrial investments in research infrastructure, and an increased potential for economic return, which is the key to getting the technology out of the laboratory and into the marketplace. Whether it is in the chemical industry, pharmaceutical industry, or others, this starts a positive feedback loop.

Michael Schrage: I am concerned about the word "important." The biggest arguments tend to be about "important" things because by definition what I do is important. Unfortunately, you may disagree with my definition. Is there anything you feel that you have learned at the Defense Advanced Research Projects Agency (DARPA) that is very good at quickly reaching consensus as to what important means?

Lawrence H. Dubois: Using DARPA as an example, what are some of the critical issues facing the safety of the nation today? I think a lot of people would say biological warfare defense is an issue we may have to deal with. You will get some consensus with that issue.

Michael Schrage: You picked the perfect example. I am going to postulate that the single most important thing that must be done for biological warfare is prevention and immunization because prophylactic cures would be administered too late to save patients. I want to overemphasize vaccination and underemphasize therapeutics. I can see a very serious debate about which approach is best. We agree that biological warfare is a big problem, but what is the tradeoff between prevention, prophylactic, and cure?

Lawrence H. Dubois: I think that is a very good question, and it certainly is open to a wider debate. Putting all your eggs in one basket is probably the wrong thing to do, however. You can look at the possibilities of success in different areas to determine where to put your money or effort, but because the probability and timing of an attack are uncertain, it is necessary to act in the shorter term.

Andrew Kaldor: But, Larry, do you have a thought on how you accomplish portfolio management in the public sector?

Lawrence H. Dubois: I think it is necessary to look at a portfolio across all of government. Again, let's use this example of biological warfare defense. There clearly has to be something in the upper right-hand corner that is of the high-reward nature. Let's forget about risk for a moment. Risk is the price paid for the high reward. The focus has to be on the reward, not on taking risk.

There is also a need to go out there and buy gas masks for people, stockpile therapeutics, and take other low-risk precautions, which would have a very high payoff if there were an attack in the near future. This doesn't mean that one agency or one organization has to cover the entire spectrum. In fact, different organizations have expertise in different areas. So it is helpful to involve an organization that knows how to do logistics, that knows how to stockpile and distribute, and that knows how to do higher-risk activities.

Francis A. Via: The concept of developing a portfolio for the public sector (government-funded research) has been extensively debated, especially within the National Science Foundation and the academic community. Yesterday it was asked whether there is something equivalent to the Semiconductor Manufacturing Technology group in the chemical industry. A number of us have participated in a technology road map development exercise for the chemical industry titled *Technology Vision 2020: The U.S. Chemical Industry*. Representatives from the three sectors of the research enterprise with an emphasis on leadership and guidance from industry participated in workshops that included representatives from government funding agencies. The workshops were held to define technical needs and develop outlines for high-risk research programs in critical technologies, including catalysts, processes, separations, analysis, measurements, engineering, and modeling.

The current focus of this activity is on the effective definition and implementation of the *Vision 2020* challenges. The Department of Energy and in particular the Office of Industrial Technology is leading this experiment for funding-focused R&D with emphasis on knowledge integration for energy savings and economic impact.

Lawrence H. Dubois: My second recommendation for portfolio management in the public sector is to focus on your goal and work backwards. Probably the most controversial is the third recommendation that entails empowering funding sources. This means giving funding sources not only the responsibility to accomplish something but also the authority to do it. For those of you who have spent time in Washington, you realize that those two things don't necessarily come together. One of the advantages of spending time at DARPA was having both the responsibility and the authority.

You need to have clear priorities, well-defined goals, and a mixed portfolio of strategic, tactical, directed, and global R&D. What is the right relationship between individual investigators and big groups? What is the right distribution between spending money on equipment and spending money on salaries? What about small science versus big science?

There are a lot of questions that will require people to think and really work the issues, as opposed to somebody in Congress making a decision based on constituent input. Ultimately, I think we have to end some programs. DARPA does it. I think other organizations need to do it as well. That is the only way we are going to start new projects because the budgets are not going to keep growing forever.

Participant: I am used to the corporate environment where I knew the rules. I had to show how I provide long-term value for my research. The thing that strikes me about funding in the public sector is that there is overspecification of what the terms and conditions are for me to do my research. It is not the money that is customized to help me do my job. I have to maneuver my job to fit some smart person's ideas of what I should be doing and how I should be doing it. This is true not only of applying for government money but also of applying for foundation funding. Everyone has his or her own idea. This is not an optimum way of doing research. I wouldn't empower the funders. I would empower the fundees, the people receiving the money.

Lawrence H. Dubois: In part, the issue depends on whether you are working in a mission-oriented agency that has a responsibility to accomplish something. Then you have got to decide what it is you want to accomplish.

Participant: No one would object to that. My objection is to how you do it.

Lawrence H. Dubois: Clearly, you have to tailor what it is you are trying to do technically with the ultimate goals of the organization. This takes somebody who is a proactive decision maker, not a passive bureaucrat.

Participant: Proactive can mean talking to the funder and helping the funder accomplish its major technical goal without saying how to do it.

Robert A. Beyerlein, National Institute of Standards and Technology: Although Larry has some very good ideas and insights on the overarching issue of empowering funding sources, I have a great skepticism about the desirability of giving those of us in Washington the responsibility to decide and implement what is important in the technical, academic, and industrial communities. I wonder if it isn't better for those in such positions to keep our ears close to the ground, to go to meetings, to rub shoulders with the technical and industrial communities, and to distill as best we can what they are telling us is important.

Lawrence H. Dubois: As a DARPA program manager, that is exactly what you are supposed to do: go out and discover what is important, find the new ideas. But there is also a need for some healthy skepticism. If you are funded today, you want your funding to continue. Of course, you are going to tell your program manager, program director, or agency that the research you are doing is by far the best and ought to be continued. There needs to be a healthy skepticism. Some people may have been doing the same thing for the past 27 years. They might have been making progress in the past, but things might have changed.

Francis A. Via: The efficiency of the funding process is a challenge that is impacting the continued progress of science. The resources consumed to secure funding have been increasing and exceed most commonly accepted estimates of 25 to 45 percent of a principal investigator's time at some of our national labs. That does not mean everybody should be funded or that we should have more money for funding, although the latter topic should be receiving greater attention. Rather, it would represent a beneficial contribution to the national research infrastructure if these workshops could more actively contribute, in some fashion, to increasing the efficiency of the funding process so that science is the winner in these processes.

Lawrence H. Dubois: In the defense sciences office at DARPA, we tried to minimize the pain and time associated with writing proposals. We put out a solicitation asking for a 4- to 6-page abstract with a basic budget describing what was planned for year 1, year 2, and year 3. Then we flagged the better abstracts and asked the authors for full proposals. This is the time-consuming part of the fund-seeking process. With this system the success rate at the full proposal level was 40 to 50 percent. This really minimized the time and effort it took people looking for funds and is now starting to be used more often.

J. Stewart Witzeman, Eastman Chemical Company: Projects that are often presented as market-driven research can, if done incorrectly, be incremental or force a silo because companies that are built on platforms of technology are no longer worrying about platforms. Instead, they are worrying about specific niches or applications and are not building the breakthrough technologies that might help them continue to be great.

While market-driven research often sounds good in practice, it can be misused or it leads companies, once they become established, to lack innovation. The question to put in front of the panel is whether there are any ways that companies can address that possibility so that they don't fall into this incrementalism.

Elsa Reichmanis: We have a broad range of activities, from very focused technology research activities that are very near to product implementation in the marketplace to very long-term fundamental research activities. Across the entire continuum of activity, we are also trying to build relationships and understanding along all aspects of the business. We aim to have a shared understanding of where the needs are and an understanding of the value of the fundamental activity between the business and research communities.

I think it boils down to having good communication systems in place for person-to-person interactions and to having communications (not the technology of it) so that there is trust and respect among all parts of the organization. This enables a silo to be avoided.

Michael Schrage: I take slight issue with that. Portfolio management is a vehicle to manage risk. How many of you have money in index funds? It should be the bulk of your investments because index funds are the safest kind. If the market goes up, you are protected. If the market goes down, you are not going to do significantly better or worse.

Innovators are inherently betting on the unbalanced portfolios. If they bet on the index, they will perform at the median within a standard deviation of how the industry group is doing.

Why are so many organizations seemingly disproportionately investing in one incremental innovation? Breakthrough innovation has become even riskier because you can call something a breakthrough, but you don't get to determine what a breakthrough is. You get to determine what a technical breakthrough is. It is your customers that determine whether it is a business breakthrough. This problem will never be solved because we can't predict the future. That is why we have a diversified portfolio.

Participant: Businesses should avoid being lock-stepped with their customers. Often they don't have very many customers, and their fortunes are tied very closely to their customers. If you do have a new idea that is not in the line of your present market, take it out of your company. Set it up separately so the project is not infected by the local culture, which doesn't allow anything beyond the norm. Move the idea to one side, set up a warehouse somewhere, and do it. Universities are another area where new ideas can be found that are worth funding. As an acquisition strategy, look for small companies for which you provide value, such as distribution. You may provide some good marketing information to them, but your new project should be a whole new technology. Get on the board of some small company and pay attention to that.

Our economy is so diverse. There are so many interesting ideas buzzing around that recognizing the good ones requires paying attention. Do not be blinded by what you have today and assume it is going to be there tomorrow.

Ned D. Heindel, Lehigh University: I wanted to make a comment about innovation with respect to the attendees at this meeting. I think if you walked around at the social hours and read our badges, you would find that the academic institutions represented among the attendees are Stevens, Youngstown, Alabama, Arkansas, Lehigh, and Maine. Are those schools Ivys? Are those places Big Tens? Are those America's leading schools? The answer is "no." Now, we do have the MIT and California Institute of Technology folks on the panel.

I would argue that in innovation and in the drive to move academic technologies to industry, the second- and third-tier universities are inherently more aggressive or at least would like to be more aggressive than the first-tier universities in this. Frankly, all of our administrators hope that there is a Gatorade or a cis-platinum in their future. Among the distribution sites I would hope we would consider for the ultimate output of this conference are the second- and third-tier academia. I don't think MIT and Harvard University need to know what we have said here. If they did know it, they wouldn't change their policies anyway.

I can tell you that when industry approaches academia with a request to create things, it is the less famous universities that are willing to give away much of the intellectual property. At Lehigh, in return for 10 fully supported graduate students working on a project on printability, adhesion, and tacticity, we're assigning the rights to a Fortune 500 chemical company. Additionally, an employee has come from that company to become a Lehigh professor, paid for by the company.

I could cite other instances of industry-friendly activities like that at the Illinois Institute of Technology and Washington State University. There are much more aggressive attempts to innovate to the mutual benefit of the company and the university in second- and third-tier schools. I hear all this grousing about the difficulty corporate representatives have had negotiating contracts with MIT. Well, come to Bethlehem, Pennsylvania, and we will make it easier for you.

Mary L. Mandich, Lucent Technologies: I have a question directed toward Michael, about the comparison of innovation and research with stocks. Maybe I am overly influenced by a book called *Stocks for the Long Run,* by Jeremy J. Siegel and Peter L. Bernstein, but the basic point of that book was that you would make the most money in the long run not if you tried to find one stock that would do superbly well, but if you backed really good horses. There is a dichotomy of market-driven research that will make a lot of money versus the fundamental belief that innovation is the way to go. I see a conflict there, and maybe you can resolve it better than I can.

Michael Schrage: The most important thing is that I am making an analogy between innovation and the stock market. They both involve investments. They are not isomorphic, but you have hit on something that was a startling revelation for me. It probably is less of a startling revelation for the people who began on the business side.

I grew up in a household believing that innovative ideas will do better in the marketplace, that innovation gets rewarded, and that people will pay a premium for innovation. They don't. People pay a premium for some innovations, but they expect other innovations to be given for free.

I thought that attitude was the exception, but in the chemical industry it turns out to be the rule. I think that one of the most painful things for people who are technically excellent to adjust to is the fact that the peer review market will pay a premium for innovation in recognition, rewards, and medals. But the marketplace where people actually pay dollars and euros will not. I believe that because so many organizations are well positioned to be fast followers, as opposed to innovators. Customers in certain market segments are counting the value of innovation. They know that if they wait a year or two, another company is going to be in the market with 80 percent of the functionality at 50 percent of the cost. Then the question is not whether we wait but how long we should wait.

This isn't true in the electronics business. I wonder whether the expectations of faster, better, and cheaper products from followers in 6 to 18 months have led to the fact that innovation becomes less valuable.

Look at market signals. It is not an accident that pharmaceutical companies are now spending more on marketing and advertising than on R&D; they want consumers to pay attention to them so they will pay a premium for these expensive drugs that have been developed.

I want to make absolutely clear that just because somebody tells you what their needs are doesn't mean they are going to pay you for satisfying them. I think that a lot of needs analysis is fundamentally misdirected. The problem with innovation is that it is not that you pay for solving the needs, but you get a premium because the risk for this is the price you pay. There has to be a premium to the risk born by the inventor.

Allen Clamen: Yes, but the market-driven research determines whether people will pay a premium for this improvement. It isn't just determining what you need and getting it to you regardless of the fact that you won't pay for it. What do you need and how valuable is it? What will it allow you to do and how much money will it save you or provide to your customer? What is the value proposition? That is part of market research.

David E. Nikles, University of Alabama: I would like to change the subject to roadblocks, but first I have a disclaimer. I fully believe in health and safety and in educating my students to be professionals, but one thing that I have worried about for the past 15 years is the intrusion of regulatory health and safety into the laboratory. I spent a month in 1990 bar code labeling all my chemicals, and then my managers wanted to get the chemicals out of my laboratory. Recently at the university, somebody said I couldn't take a chemical from my laboratory out of the chemistry building because it could be used as a terrorist weapon.

Do we want our scientists bogged down with all of this intrusion of regulation into the innovation process? I think everyone would react by thinking, "Of course, we have to do it," but such regulation does not exist worldwide.

When I raise these issues, it is heresy to our health and safety establishment at the university. Are we going to have American scientists tied up, dealing with the audit trails that show that they are safe and limiting their ability to innovate? Sooner or later there will be a rule that prohibits chemicals in the laboratory.

Allen Clamen: In the corporate environment we have been living with that all along. You are talking about the university catching up to the corporate environment. That is not such a bad thing because many of those students will end up in industry.

David E. Nikles: I agree with that, but at some point do the regulations become so onerous that I am losing productivity and the benefit of the research?

Elsa Reichmanis: I don't really find that the regulations are so very onerous from my perspective in a research lab in industry. Now, we do have an organization in place that works with those issues and helps us meet the regulations. However, I think it is very important to work in a safe environment both for the individual and for the entire community. I would hate to be in a situation where the pollution level is through the roof compared to the United States, and hence the quality of life is lower. We are not going to have a safe environment without regulations.

Venkat Venkatasubramanian: Half of my work involves in-process safety and monitoring of chemical plants to prevent Union Carbide kinds of incidents. Other countries are also catching up. Certainly, in India, for example, the number of regulations greatly increased after the Bopal incident, and people are much more concerned and aware of health, safety, and environmental issues. As society evolves, life is of increasingly higher value, and that is reflected in regulations.

Participant: If you really want to get ahead of the game, invite DuPont in. DuPont is number 1 in safety, and its costs are lower because of the serious attention the company pays to safety.

David J. Soderberg, BP Chemicals: I would like to give an industrial perspective on that. First, I will not do any work with any university that I consider to be unsafe for a very practical reason. If there is a liability inherent in the work being done, I do not want to be responsible financially or otherwise for the impact. In fact, when university professors and students come in, we give them training in our safety procedures. When we send people out to collaborators' laboratories, our employees will actually perform a safety audit in that lab. That is very important.

Second, I think it is a misconception to say that developing countries lack standards. That might be from some local industry and historical perspective, but certainly if you are putting a new plant on the ground in Taiwan, the authorities will expect you to comply with TA-Luft, the strict German Technical Directive for Air Pollution Abatement, which is among the toughest air quality legislation that industry has to comply with. It is actually tougher in some ways to establish foreign plants than some domestic plants.

Participant: I want to make a comment on this particular issue because I worked in a national laboratory that changed from virtually unregulated to very tightly regulated between the mid-1980s and the late 1990s. Then I left the national laboratory and worked for DuPont for a little over a year, where safety was an important consideration. I observed that moving from an unregulated environment where this is not a consideration to a regulated environment where health and safety are important, there is a period of time during which this is disruptive, because you don't have the mechanisms and the support structures in place to help you deal with it.

People are very averse to change. This is human nature. When they are told to do something differently, their immediate response is to find all of the problems with the new method. Once people become used to the new methodology, once some of the legacy problems are dealt with, and once a support structure is in place, it becomes much less burdensome. I saw that occur. Initially, anything could be poured down the sink. Then a paper towel with a small amount of acetone on it couldn't be thrown in the trash after the acetone evaporated. Finally, regulations came back to a rational point and could be dealt with.

The reality was that my work procedures in the national laboratory were not very different than the procedures at DuPont. DuPont did have a better infrastructure to help employees deal with regulated procedures. Therefore, I do not believe that health and safety issues are a burden on research. I think most industrial people who have worked in manufacturing find that in the end they actually help because of all the different ways they make processes more efficient and liability lower.

Venkat Venkatasubramanian: I would like to comment on structured invention and innovation linked to reduced cycle time. I want to draw a distinction between the two kinds of inventions or discoveries, what I call structural and parametric changes. For example, you can go from A to B by car or by plane. These are two completely different modes of transportation. I would call them structural changes. The invention of the plane would be a structural change. Once the Wright Brothers created their plane, coming up with the Boeing 747 in my view is a parametric change. That is, you have the basic structure, but you try to optimize it for speed, efficiency, and so on.

Development of road maps or structured innovation plans would probably work for parametric improvements around the structure, but the basic structure itself is unpredictable and cannot be regulated because the future cannot be predicted. Road maps can plan the incremental improvements, which lead to important changes, but they cannot plan breakthrough or leap frog types of discoveries.

Appendixes

A

Workshop Participants

Richard C. Alkire, University of Illinois at Urbana-Champaign
William L. Alworth, Tulane University
Robert A. Beyerlein, National Institute of Standards and Technology
Robert J. Bianchini, Colgate-Palmolive Company
Christopher S. Brazel, University of Alabama
Kimberly A. Brown, United States Army
Leonard J. Buckley, Defense Advanced Research Projects Agency
Thomas W. Chapman, National Science Foundation
Allen Clamen, ExxonMobil (retired)
David L. Cocke, Lamar University
Richard J. Colton, Naval Research Laboratory
Walter G. Copan, The Lubrizol Corporation
Lawrence H. Dubois, SRI International
Joseph S. Francisco, Purdue University
Brian G. Frederick, University of Maine
Gregory S. Girolami, University of Illinois at Urbana-Champaign
Louis C. Glasgow, E. I. du Pont de Nemours and Company
Mary L. Good, University of Arkansas, Little Rock
Richard M. Gross, The Dow Chemical Company
James R. Heath, University of California, Los Angeles
L. Louis Hegedus, ATOFINA Chemicals, Inc.
Ned D. Heindel, Lehigh University
Michael J. Holland, Office of Science and Technology Policy
Robert W. R. Humphreys, National Starch and Chemical Company
Andrew Kaldor, ExxonMobil
William H. Kirchhoff, U.S. Department of Energy
Richard K. Koehn, Salus Therapeutics, Inc.

Peter A. Koen, Stevens Institute of Technology
Russell Koveal, ExxonMobil
Theresa Laranang-Mutlu, American Chemical Society
Daniel P. Leta, ExxonMobil
Flint Lewis, American Chemical Society
Mary L. Mandich, Lucent Technologies
James H. Mike, Youngstown State University
Daryl W. Mincey, Youngstown State University
David E. Nikles, University of Alabama
Parry M. Norling, RAND
Nancy L. Parenteau, Organogenesis, Inc.
Kenneth A. Pickar, California Institute of Technology
David R. Rea, E. I. du Pont de Nemours and Company (retired)
Elsa Reichmanis, Lucent Technologies
C. E. Rinehart, ExxonMobil
Michael Schrage, Massachusetts Institute of Technology
William G. Schulz, American Chemical Society
David J. Soderberg, BP Chemicals
Guangyu Sun, Georgetown University
Hans Thomann, ExxonMobil
Kimberly W. Thomas, Los Alamos National Laboratory
Yuye Tong, Georgetown University
Venkat Venkatasubramanian, Purdue University
Francis A. Via, Fairfield Resources, Inc.
Henry F. Whalen, Jr., American Chemical Society
J. Stewart Witzeman, Eastman Chemical Company
Matthew Wolfe, National Institutes of Health

Appendix B

Biographical Sketches of Workshop Speakers

Allen Clamen (now retired) was senior advisor for marketing/technology value creation at ExxonMobil Chemical Company in Houston, Texas. He was responsible for developing effective and efficient processes for idea management, portfolio management, and stage gating of new product development projects. The latter process has been used successfully for managing the translation of ideas from conception to commercial reality for over 10 years. Since 1998 this management system has been enhanced by the incorporation of a portfolio management process allowing businesses to evaluate multiple opportunities on a consistent, objective basis.

Starting with Exxon Research and Engineering Company (Linden, NJ) in 1966 after graduating with a Ph.D. in chemical engineering from McGill University, Clamen occupied several positions before transferring to Exxon Chemical in 1976, where he served as halobutyl operations manager (Baton Rouge, LA), butyl manufacturing manager (Baytown, TX), and polymers advanced development manager (Baytown, TX). In 1986 he returned to Linden, NJ, as butyl polymers technology manager and polymers site manager before returning to Baytown, TX, in 1994 to assume the position of technology processes manager. For the past 4 years he has led teams to create value via improved marketing and technology processes.

Lawrence H. Dubois received his B.S. degree in chemistry from the Massachusetts Institute of Technology in 1976 and a Ph.D. in physical chemistry from the University of California, Berkeley, in 1980. Dubois then joined AT&T Bell Laboratories in Murray Hill, NJ, to pursue studies of the chemistry and physics of metal, semiconductor, and insulator surfaces; chemisorption and catalysis by materials formed at the metal-semiconductor interface; and novel methods of materials growth and preparation.

In 1987 he was promoted to distinguished member of the technical staff and technical manager. His efforts broadened to include projects on polymer-surface interactions; adhesion promotion; corrosion protection; chemical vapor deposition and thin film growth; optical fiber coating; synthesis, structure, and reactivity of model organic surfaces; and time-resolved surface vibrational spectroscopy.

In 1993, Dubois moved to the Massachusetts Institute of Technology Lincoln Laboratory as a senior staff scientist and was assigned to the Defense Advanced Research Projects Agency (DARPA). In that capacity, he established the Advanced Energy and Environmental Technologies Program and managed projects on the development and manufacturing of rechargeable batteries; high-performance, direct methanol, and logistic fuel-powered fuel cells; and the development of new, more environmentally sound manufacturing processes, environmental sensors, and waste destruction/reclamation procedures.

In 1995, Dubois was promoted to deputy director and in 1996 to director of the Defense Sciences Office at DARPA. This office is responsible for an annual investment of approximately $300 million toward the development of technologies for biological warfare defense, biology, defense applications of advanced mathematics, and materials and devices for new military capabilities.

In March 2000, Dubois joined SRI International as vice president and head of the Physical Sciences Division, a group of over 150 scientists and engineers focusing on the development and commercialization of advanced materials, microfabrication technologies, power sources, biological warfare defense, medical diagnostics, molecular and optical physics, explosives and propellants, catalysts, coatings, and environmentally benign processing.

Dubois is the author of over 130 publications and holds four U.S. and several foreign patents. His numerous honors include the prestigious IR100 and Alpha Chi Sigma awards as well as the Office of the Secretary of Defense Award for Outstanding Achievement and the Secretary of Defense Medal for Outstanding Public Service. He sits on the Board of Directors of two spin-off companies from SRI: Polyfuel and CYANCE.

Mary L. Good is the Donaghey University Professor at the University of Arkansas, Little Rock, and serves as the managing member for Venture Capital Investors, LLC, a group of Arkansas business leaders who expect to foster economic growth in the area through the opportunistic support of technology-based enterprises. Good also presently serves on the Board of Biogen, a successful biotech company in Cambridge, Massachusetts; IDEXX Laboratories of Westbrook, Maine; and the Lockheed Martin Energy Research Corporation Board of Oak Ridge, Tennessee. Previously Good served 4 years as the under secretary for technology for the Technology Administration in the Department of Commerce, a presidentially appointed, Senate-confirmed position.

The Technology Administration is the focal point in the federal government for assisting U.S. Industry to improve its productivity, technology, and innovation in order to compete more effectively in global markets. In particular, the administration works with industry to eliminate legislative and regulatory barriers to technology commercialization and to encourage adoption of modern technology management practices. The Technology Administration is comprised of the National Institute of Standards and Technology, the National Technical Information Service, the Office of the Assistant Secretary for Technology Policy, and the Office of Air and Space Commercialization.

In addition to her role as under secretary for technology, Good chaired the National Science and Technology Council's Committee on Technological Innovation, and served on the Council's Committee on National Security.

Before joining the administration, Good was senior vice-president of technology at Allied Signal, Inc., where she was responsible for the centralized research and technology organizations with facilities in Morristown, NJ; Buffalo, NY; and Des Plaines, IL. She was a member of the Management Committee and was responsible for technology transfer and commercialization support for new technologies. This position followed assignments as president of Allied Signal's Engineered Materials Research Center, director of the UOP Research Center, and president of the Signal Research Center. Good's accomplishments in industrial research management are the achievements of a second career, having moved to an industrial position after more than 25 years of teaching and research in the Louisiana State University system. Before joining Allied Signal, she was professor of chemistry at the University of

New Orleans and professor of materials science at Louisiana State University, where she achieved the university's highest professional rank, Boyd Professor.

Good was appointed to the National Science Board by President Carter in 1980 and again by President Reagan in 1986. She was chairman of that board from 1988 until 1991, when she received an appointment from President Bush to become a member of the President's Council of Advisors on Science and Technology (PCAST). Good also served on the boards of Rensselaer Polytechnic Institute, Cincinnati Milacron, and Ameritech. She was also a member of the National Advisory Board for the State of Arkansas.

Good is an elected member of the National Academy of Engineering, a past president of the American Chemical Society, a fellow of the American Association for the Advancement of Science, and a member of the American Institute of Chemists and the Royal Society of Chemistry. She has been active on the boards of directors of such groups as the Industrial Research Institute, Oak Ridge Associated Universities, and the National Institute for Petroleum and Energy Research. She has also served on advisory panels for the National Research Council, the National Bureau of Standards, the National Science Foundation Chemistry Section, the National Institutes of Health, and NASA, and on the executive committee for the International Union of Pure and Applied Chemistry.

Good received the National Science Foundation's Distinguished Public Service Award, the Albert Fox Demers Medal Award from Rensselaer Polytechnic Institute, the American Association for the Advancement of Science Award, and the American Institute of Chemists' Gold Medal and was chosen Scientist of the Year by *Industrial Research and Development* magazine. She was elected a foreign member of the Royal Swedish Academy of Engineering Sciences in 1990, became a member of the Tau Beta Pi Association (the Engineering Honor Society), was awarded the Charles Lathrop Parsons Award of the American Chemical Society, and received the Industrial Research Institute Medalist Award. In 1997 she received the Priestly Medal from the American Chemical Society, the highest award given by the society. She has published over 100 articles in reference journals and is the author of a book, *Integrated Laboratory Sequence*, published by Barnes and Noble.

Good received her B.S. in chemistry from the University of Central Arkansas and her M.S. and Ph.D. degrees in inorganic chemistry from the University of Arkansas. She also received numerous awards and honorary degrees from many colleges and universities, including most recently the College of William and Mary, Polytechnic University of New York, Louisiana State University, and Michigan State University.

Richard M. Gross is corporate vice president of research and development for the Dow Chemical Company. In this capacity he serves on Dow's Corporate Operating Board, Human Resources Committee, Retirement Board, and Corporate Contributions Committee.

Until March 1998 Gross was vice president and director of the Michigan operations and global vice president of Core Technologies R&D. Gross joined Dow in 1974 in the hydrocarbons and energy research area of the Michigan division R&D. In 1979 he transferred to the Louisiana division, where he spent several years in coal gasification research before returning to Midland. Gross was technical director for Consumer Products Research in Dow's Michigan Applied Science and Technology Laboratories. He was also director of R&D and Technical Services and Development for Chemicals and Metals. In 1992 he was named R&D director for North American Chemicals and Metals/Hydrocarbons R&D.

Gross was a 1996 recipient of the Dow Genesis Award for Excellence in People Development. He is a member of the American Chemical Society, the American Institute of Chemical Engineers, the Industrial Research Institute, and the Council for Chemical Research, where he serves on the Governing

Board's Executive Committee and recently was elected to the office of first vice chair, the Chemical Engineering Advisory Board at Worcester Polytechnic Institute, the Advisory Board of the National Science Resources Center, the Advisory Board for the College of Chemistry at the University of California, Berkeley, the National Research Council's Board on Chemical Sciences and Technology and the Michigan Molecular Institute Board.

James R. Heath received a B.Sc. degree in chemistry from Baylor University in 1984 and a Ph.D. degree in chemistry from Rice University in 1988, where he studied in the group of Richard E. Smalley. Heath was a Miller postdoctoral fellow at UC Berkeley from 1988 to 1991, where he worked in the laboratory of Richard J. Saykally. He was a research staff member at the IBM T. J. Watson Research Laboratories in Yorktown Heights, New York, from 1991 to 1994. In 1994 he left IBM to join the Department of Chemistry and Biochemistry at UCLA. He was promoted to tenure in 1996 and to full professor in 1997. He is currently the director of the California NanoSystems Institute, which was formed by California Governor Grey Davis in December 2000. Heath was a David and Lucile Packard Fellow (1994 to 1999) and an Alfred P. Sloan Fellow (1997). He is a fellow of the American Physical Society and has received the Jules Springer Award in Applied Physics (2000), the Feynman Prize (2000), and the Sackler Prize in the Physical Sciences (2001). Heath's research interests focus on "artificial" quantum dot solids and quantum phase transitions in those solids; molecular electronics architecture, devices, and circuitry; and the spectroscopy and imaging of transmembrane proteins in physiological environments.

Richard K. Koehn, the author of more than 100 papers and co-editor of *The Evolution of Genes and Proteins*, was professor of ecology and evolution at the State University of New York at Stony Brook (1970 to 1992) where he was also dean of biological sciences (1978 to 1988) and director of the Center for Advanced Biomedical Biotechnology for New York State (1983 to 1992). Koehn was vice president for research at the University of Utah from 1992 to August 2000. He is currently president and CEO of Salus Therapeutics, Inc., an emerging biotechnology company in Salt Lake City, Utah.

Koehn has been a member of the boards of directors of several organizations, including the Council on Biotechnology, the Association of Biotechnology Companies, the Long Island Forum for Technology, the Long Island High Technology Incubator Management Corporation, the New York Biotechnology Association, the Boyce Thompson Institute for Plant Research, the Organization of Tropical Studies, the Advisory Council to the Vice-Chair of the New York Legislative Commission on Science and Technology, and the Commission on Biomedical Research of the New York Academy of Medicine. In Utah he served 8 years on the Governor's Council on Science and Technology and as president of the University of Utah Research Foundation, Inc. He is past chair of the board of trustees of the Association of Western Universities, a member of the executive committee of the Council on Research Policy and Graduate Education of NASULGC, and a director of the Utah Life Sciences Industry Association. He is currently a member of the Investment Advisory Board of Utah Ventures II and a director of the Alberta Henry Educational Foundation and Ballet West.

Koehn has lectured on evolutionary genetics, biotechnology policy, entrepreneurial universities, and the responsible conduct of scientific research in more than 20 countries. He has been a Guggenheim fellow, a NATO senior science fellow, and the recipient of a number of awards for leadership in the New York biotechnology industry. In 1991 he was recipient of the Entrepreneur of the Year Award from Ernst & Young/Merrill Lynch/ *Inc. Magazine.* In 1997 he was awarded a gold Aurora for the production of *Learning Through Discovery*, a local public television series on student involvement in research. In 2001 he was recognized as a distinguished alumnus by the College of Liberal Arts and Sciences of Arizona State University.

Kenneth A. Pickar earned a Ph.D. degree in low-temperature physics at the University of Pennsylvania. He joined Bell Laboratories in Murray Hill, NJ, where he worked in the areas of ion implantation and electron beam technology. He has 50 publications and talks on these subjects, including a review of beam processing, written while on a sabbatical year at the Technion-Israel Institute of Technology. At GE Corporate R&D he was responsible for all electronics research from semiconductor materials through large medical imaging systems, lighting, radar, etc. At AlliedSignal Corporation he was senior vice president for engineering and technology and chairman of the Corporate Technology Board. His responsibilities in aerospace technologies included jet engines, aircraft braking, collision avoidance, managing NASA ground stations, and the like. He was the "champion" for the Aerospace Product Development Process, leading the development of "faster, better, cheaper" ways of developing new products. In 1998 he became visiting professor of mechanical engineering at Caltech, where he teaches courses on the engineering design of products and the management of technology. In 1999 he was named the J. Stanley Johnson Professor at Caltech. He is co-principal investigator of the National Science Foundation-funded Entrepreneurial Fellowship Program. Pickar has served on a number of university advisory committees, including Stanford, Berkeley, Cornell, and Illinois and the Technical Advisory Committee of the Council on Competitiveness. He was vice chairman of the Microelectronics and Computer Consortium, on the Board of Directors of the Semiconductor Research Corporation, and a director of the Albany Medical Center, Level One Corporation, and NeuStar Corporation.

Elsa Reichmanis, director of materials research department at Bell Laboratories, Lucent Technologies in Murray Hill, NJ, is president of the American Chemical Society, the world's largest scientific society. Her presidency began January 1, 2003.

Only the fourth woman to be elected president in the society's 125-year history, Reichmanis pledges to work with sister societies to increase the visibility of science and technology to government leaders. "Our elected officials must understand the importance of the chemical workforce's contributions to the health and welfare of our nation," she says. "Today's fundamental research efforts are the building blocks for tomorrow's breakthroughs and innovations."

An ACS member since 1973, Reichmanis is also a member of the American Physical Society, the Institute of Electrical and Electronics Engineers, the American Association for the Advancement of Science, the Materials Research Society, the National Academy of Engineering, and the Society of Women Engineers. She recently chaired the ACS Committee on Science and is an associate editor of the ACS journal *Chemistry of Materials*. She serves on the U.S. National Committee for the International Union of Pure and Applied Chemistry and the National Research Council Panel for Materials and Science Engineering.

Among Reichmanis's many honors is the Society of Chemical Industry's 2001 Perkin Medal for her pioneering contributions to designing materials that allow silicon chips to continue shrinking in size while also improving in performance. She has authored more than 120 journal articles, edited five books, and organized and chaired numerous national and international symposia. She holds 17 U.S. patents. Reichmanis earned her B.S. (1972) and Ph.D. (1975) at Syracuse University. She resides in Westfield, NJ.

Michael Schrage is co-director of the Massachusetts Institute of Technology Media Lab's eMarkets Initiative and a senior advisor to the MIT Security Studies Program. His research focuses on the role of models, prototypes, and simulations as essential media for managing innovation and risk. His book, *Serious Play* (Harvard Business School Press, 2000), explores the economics and ethology of modeling within organizations.

In addition, he is executive director of the Merrill Lynch Innovation Grants Competition, which rewards doctoral students who present novel approaches to commercializing their thesis research. He serves on Ticketmaster's Board of Directors and on the editorial board of the *Sloan Management Review*. He is a consultant to such companies as British Petroleum, MasterCard, Millennium Pharmaceutical, and Bosch. He currently writes a column on innovation issues for *Technology Review* and on implementation issues for *CIO* magazine. Schrage has also been a contributor to the *Harvard Business Review*, the *Washington Post*, the *Wall Street Journal,* and *Wired* magazine.

Venkat Venkatasubramanian is university faculty scholar professor of chemical engineering at Purdue University. He received his Ph.D. in chemical engineering from Cornell University in 1984, his M.S. in physics from Vanderbilt University in 1979, and his B.Tech. in chemical engineering from the University of Madras, India, in 1977. Venkat worked as a research associate in artificial intelligence in the Department of Computer Science at Carnegie-Mellon University and taught at Columbia University before joining Purdue in 1988. At Purdue, Venkat directs the research efforts of several graduate students and co-workers in the Laboratory for Intelligent Process Systems. Venkatasubramanian's research contributions have been in the areas of process fault diagnosis and supervisory control, hazard and safety analysis, operating procedures synthesis for batch processes, product formulation and design, complex adaptive systems, using knowledge-based systems, neural networks, genetic algorithms, mathematical programming, and statistical approaches. His teaching interests include courses in artificial intelligence, process design and control, statistical thermodynamics, and applied statistics.

Venkatasubramanian has published over 110 refereed papers and has delivered over 100 invited lectures and seminars, including six keynote lectures, at various international conferences and institutions all over the world. He has authored a three-volume CACHE case study, *Knowledge-Based Systems for Heuristic Classification Problems in Process Engineering,* and has co-authored a monograph, *Advanced Knowledge Representation.* Venkat has chaired or co-chaired over 30 international meetings, conferences, and sessions in the areas of artificial intelligence applications in process engineering. Fourteen doctoral and five master's students have graduated under Venkat's supervision. Venkat has been a consultant to several major global corporations and institutions such as Air Products, ALCOA, American Cynamid, Arthur D. Little, Amoco, Caterpillar, DowAgro Sciences, Exxon, Honeywell, Lubrizol, the United Nations (UNIDO and UNDP), Indian Oil, ICI (U.K.), Nova Chemicals, and G.D. Searle.

Venkatasubramanian's contributions have been recognized by several awards and honors. He was the 1990 recipient of the Eminent Overseas Lectureship Award from the Institution of Engineers in Australia. He was a guest co-editor of the special issue of *Computers and Chemical Engineering* on *Neural Networks* in 1992. In 1993 he was awarded the United Nations Development Program Invited Lectureship at the Indian Institute of Technology in Delhi, India. He received the Norris Shreve Award for Outstanding Teaching in Chemical Engineering in 1993. He is an academic trustee and vice president of the Computer Aids for Chemical Engineering Corporation, a nonprofit organization for the promotion of computers in chemical engineering education. He also served on the editorial board of the *Process Safety Progress* journal published by the American Institute of Chemical Engineers. He currently serves on the editorial board of *Computers and Chemical Engineering.* His recent paper on fault diagnosis was awarded the CAST Directors' Award for the Best Poster Presentation at the AIChE annual meeting in Los Angeles in 2000. In 1996, based on Venkatasubramanian's research contributions, *Industry Week* magazine selected him as "one of the fifty R&D stars in the United States whose achievements are shaping the future of our industrial culture and America's technology policy."

Francis A. Via joined Fairfield Resources International as a senior consultant after more than 30 years managing industrial R&D, intellectual property, and market development at Stauffer Chemical Company, Akzo Nobel, Inc., and GE. He achieved more than a dozen commercial successes, yielding hundreds of millions of dollars in new markets or savings in specialty chemicals, catalysts, agricultural chemicals, pharmaceutical intermediates, and polymers.

Via began his career with Stauffer Chemical Company in 1970 and, with its acquisition by Akzo Nobel in 1987, became part of the R&D leadership team. He directed Akzo Nobel's Corporate Research-US to capture emerging technology in catalysis, advanced materials, electronic chemicals, immuno-diagnostics, and biochemistry. Utilizing external cooperative research programs at universities and national laboratories served as the keystone for this corporate research. In 1998 he accepted a challenge to build and manage a catalyst research group at the GE Corporate R&D Center to develop fuel cells, carbonylation catalysis, combinatorial chemistry for catalysis, process technology, and medical imaging agents.

Via is the recipient of numerous awards, including two internal GE awards for technical excellence and productivity. In 1994 he was elected Fellow of the American Association for the Advancement of Science. He is the recipient the first Department of Energy Office of Industrial Technology's Industrial Partnership Award (1999). He authored a chapter for the American Chemical Society's millennium publication *Chemical Research—2000 & Beyond*. In addition, he serves as a consultant or on review committees for the Department of Energy, the National Science Foundation, the National Academy of Sciences and the American Chemical Society. In 1999, Via was selected to the Chemical Industry Executive Steering Group for the DOE-OIT.

Via has 24 patents, 11 publications, and more than 25 invited presentations. He holds a B.S. degree from West Virginia University and a Ph.D. from Ohio State. He attended management training at the Wharton School, Polytechnic U, and other programs. He is an active member of the Council for Chemical Research, served on its Board of Directors from 1995 to 1997 and chaired the Science Education Committee from 1993 to 1995. He chaired the External Research Director's Network from 1995 to 1997 for the Industrial Research Committee.

Appendix C

Origin of and Information on the Chemical Sciences Roundtable

In April 1994 the American Chemical Society (ACS) held an Interactive Presidential Colloquium entitled "Shaping the Future: The Chemical Research Environment in the Next Century."[1] The report from this colloquium identified several objectives, including the need to ensure communication on key issues among government, industry, and university representatives. The rapidly changing environment in the United States for science and technology has created a number of stresses on the chemical enterprise. The stresses are particularly important with regard to the chemical industry, which is a major segment of U.S. industry, makes a strong, positive contribution to the U.S. balance of trade, and provides major employment opportunities for a technical work force. A neutral and credible forum for communication among all segments of the enterprise could enhance the future well-being of chemical science and technology.

After the report was issued, a formal request for such a roundtable activity was transmitted to Bruce M. Alberts, chairman of the National Research Council (NRC), by the Federal Interagency Chemistry Representatives, an informal organization of representatives from the various federal agencies that support chemical research. As part of the NRC, the Board on Chemical Sciences and Technology (BCST) can provide an intellectual focus on issues and fundamentals of science and technology across the broad fields of chemistry and chemical engineering. In the winter of 1996, Dr. Alberts asked BCST to establish the Chemical Sciences Roundtable to provide a mechanism for initiating and maintaining the dialogue envisioned in the ACS report.

The mission of the Chemical Sciences Roundtable is to provide a science-oriented, apolitical forum to enhance understanding of the critical issues in chemical science and technology affecting the government, industrial, and academic sectors. To support this mission, the Chemical Sciences Roundtable will do the following:

---

[1] *Shaping the Future: The Chemical Research Environment in the Next Century,* American Chemical Society Report from the Interactive Presidential Colloquium, April 7-9, 1994, Washington, DC.

APPENDIX C								129

• Identify topics of importance to the chemical science and technology community by holding periodic discussions and presentations and gathering input from the broadest possible set of constituencies involved in chemical science and technology.

• Organize workshops and symposia and publish reports on topics important to the continuing health and advancement of chemical science and technology.

• Disseminate the information and knowledge gained in the workshops and reports to the chemical science and technology community through discussions with, presentations to, and engagement of other forums and organizations.

• Bring topics deserving further in-depth study to the attention of the NRC's Board on Chemical Sciences and Technology. The roundtable itself will not attempt to resolve the issues and problems that it identifies–it will make no recommendations, nor provide any specific guidance. Rather, the goal of the roundtable is to ensure a full and meaningful discussion of the identified topics so that the participants in the workshops and the community as a whole can determine the best courses of action.